A Practical Guide to Scientific Data Analysis

A Practical Guide to Scientific Data Analysis

David Livingstone
ChemQuest, Sandown, Isle of Wight, UK

A John Wiley and Sons, Ltd., Publication

This edition first published 2009
© 2009 John Wiley & Sons, Ltd

Registered office

John Wiley & Sons Ltd, The Atrium, Southern Gate, Chichester, West Sussex, PO19 8SQ, United Kingdom

For details of our global editorial offices, for customer services and for information about how to apply for permission to reuse the copyright material in this book please see our website at www.wiley.com.

The right of the author to be identified as the author of this work has been asserted in accordance with the Copyright, Designs and Patents Act 1988.

Reprinted August 2010

Library of Congress Cataloging-in-Publication Data

Livingstone, D. (David)
 A practical guide to scientific data analysis / David Livingstone.
  p. cm.
 Includes bibliographical references and index.
 ISBN 978-0-470-85153-1 (cloth : alk. paper)
 1. QSAR (Biochemistry) – Statistical methods. 2. Biochemistry – Statistical methods.
I. Title.
 QP517.S85L554 2009
 615′.1900727–dc22

                                                           2009025910

A catalogue record for this book is available from the British Library.

ISBN 978-0470-851531

Typeset in 10.5/13pt Sabon by Aptara Inc., New Delhi, India.
Printed and bound in Great Britain by TJ International, Padstow, Corwall

This book is dedicated to the memory of my first wife, Cherry (18/5/52–1/8/05), who inspired me, encouraged me and helped me in everything I've done, and to the memory of Rifleman Jamie Gunn (4/8/87–25/2/09), whom we both loved very much and who was killed in action in Helmand province, Afghanistan.

Contents

Preface

The idea for this book came in part from teaching quantitative drug design to B.Sc. and M.Sc. students at the Universities of Sussex and Portsmouth. I have also needed to describe a number of mathematical and statistical methods to my friends and colleagues in medicinal (and physical) chemistry, biochemistry, and pharmacology departments at Wellcome Research and SmithKline Beecham Pharmaceuticals. I have looked for a textbook which I could recommend which gives *practical* guidance in the use and interpretation of the apparently esoteric methods of multivariate statistics, otherwise known as pattern recognition. I would have found such a book useful when I was learning the trade, and so this is intended to be that sort of guide.

There are, of course, many fine textbooks of statistics and these are referred to as appropriate for further reading. However, I feel that there isn't a book which gives a practical guide for scientists to the processes of data analysis. The emphasis here is on the application of the techniques and the interpretation of their results, although a certain amount of theory is required in order to explain the methods. This is not intended to be a statistical textbook, indeed an elementary knowledge of statistics is assumed of the reader, but is meant to be a statistical companion to the novice or casual user.

It is necessary here to consider the type of research which these methods may be used for. Historically, techniques for building models to relate biological properties to chemical structure have been developed in pharmaceutical and agrochemical research. Many of the examples used in this text are derived from these fields of work. There is no reason, however, why any sort of property which depends on chemical structure should not be modelled in this way. This might be termed quantitative structure–property relationships (QSPR) rather than QSAR where

A stands for activity. Such models are beginning to be reported; recent examples include applications in the design of dyestuffs, cosmetics, egg-white substitutes, artificial sweeteners, cheese-making, and prepared food products. I have tried to incorporate some of these applications to illustrate the methods, as well as the more traditional examples of QSAR.

There are also many other areas of science which can benefit from the application of statistical and mathematical methods to an examination of their data, particularly multivariate techniques. I hope that scientists from these other disciplines will be able to see how such approaches can be of use in their own work.

The chapters are ordered in a logical sequence, the sequence in which data analysis might be carried out – from planning an experiment through examining and displaying the data to constructing quantitative models. However, each chapter is intended to stand alone so that casual users can refer to the section that is most appropriate to their problem. The one exception to this is the Introduction which explains many of the terms which are used later in the book. Finally, I have included definitions and descriptions of some of the chemical properties and biological terms used in panels separated from the rest of the text. Thus, a reader who is already familiar with such concepts should be able to read the book without undue interruption.

David Livingstone
Sandown, Isle of Wight
May 2009

Abbreviations

π	hydrophobicity substituent constant
σ	electronic substituent constant
Λ_{alk}	hydrogen-bonding capability parameter
ΔH	enthalpy
AI	artificial intelligence
ANN	artificial neural networks
ANOVA	analysis of variance
BPN	back-propagation neural network
CA	cluster analysis
CAMEO	Computer Assisted Mechanistic Evaluation of Organic reactions
CASE	Computer Assisted Structure Evaluation
CCA	canonical correlation analysis
CoMFA	Comparative Molecular Field Analysis
CONCORD	CONnection table to CoORDinates
CR	continuum regression
CSA	cluster significance analysis
DEREK	Deductive Estimation of Risk from Existing Knowledge
ED_{50}	dose to give 50 % effect
ESDL10	electrophilic superdelocalizability
ESS	explained sum of squares
FA	factor analysis
FOSSIL	Frame Orientated System for Spectroscopic Inductive Learning
GABA	γ-aminobutyric acid
GC-MS	gas chromatography-mass spectrometry
HOMO	highest occupied molecular orbital
HPLC	high-performance liquid chromatography

HTS	high throughput screening
I_{50}	concentration for 50 % inhibition
IC_{50}	concentration for 50 % inhibition
ID3	iterative dichotomizer three
IR	infrared
K_m	Michaelis–Menten constant
KNN	k-nearest neighbour technique
LC_{50}	concentration for 50 % lethal effect
LD_{50}	dose for 50 % death
LDA	linear discriminant analysis
LLM	linear learning machine
$\log P$	logarithm of a partition coefficient
LOO	leave one out at a time
LV	latent variable
m.p.	melting point
MAO	monoamine oxidase
MIC	minimum inhibitory concentration
MLR	multiple linear regression
mol.wt.	molecular weight
MR	molar refractivity
MSD	mean squared distance
MSE	explained mean square
MSR	residual mean square
MTC	minimum threshold concentration
NLM	nonlinear mapping
NMR	nuclear magnetic resonance
NOA	natural orange aroma
NTP	National Toxicology Program
OLS	ordinary least square
PC	principal component
PCA	principal component analysis
PCR	principal component regression
p.d.f.	probability density function
pI_{50}	negative log of the concentration for 50 % inhibition
PLS	partial least squares
PRESS	predicted residual sum of squares
QDA	quantitative descriptive analysis
QSAR	quantitative structure-activity relationship
QSPR	quantitative structure-property relationship
R^2	multiple correlation coefficient
ReNDeR	Reversible Non-linear Dimension Reduction

RMSEP	root mean square error of prediction
RSS	residual or unexplained sum of squares
SE	standard error
SAR	structure-activity relationships
SIMCA	see footnote p. 195
SMA	spectral map analysis
SMILES	Simplified Molecular Input Line Entry System
SOM	self organising map
TD_{50}	dose for 50 % toxic effect
TOPKAT	Toxicity Prediction by Komputer Assisted Technology
TS	taboo search
TSD	total squared distance
TSS	total sum of squares
UFS	unsupervised forward selection
UHTS	ultra high throughput screening
UV	ultraviolet spectrophotometry
V_m	Van der Waals' volume

1

Introduction: Data and Its Properties, Analytical Methods and Jargon

Points covered in this chapter

- Types of data
- Sources of data
- The nature of data
- Scales of measurement
- Data distribution
- Population and sample properties
- Outliers
- Terminology

PREAMBLE

This book is not a textbook although it does aim to teach the reader how to do things and explain how or why they work. It can be thought of as a handbook of data analysis; a sort of workshop manual for the mathematical and statistical procedures which scientists may use in order to extract information from their experimental data. It is written for scientists who want to analyse their data 'properly' but who don't have the time or inclination to complete a degree course in statistics in order

A Practical Guide to Scientific Data Analysis   David Livingstone
© 2009 John Wiley & Sons, Ltd

to do this. I have tried to keep the mathematical and statistical theory to a minimum, sufficient to explain the basis of the methods but not too much to obscure the point of applying the procedures in the first case.

I am a chemist by training and a 'drug designer' by profession so it is inevitable that many examples will be chemical and also from the field of molecular design. One term that may often appear is QSAR. This stands for Quantitative Structure Activity Relationships, a term which covers methods by which the biological activity of chemicals is related to their chemical structure. I have tried to include applications from other branches of science but I hope that the structure of the book and the way that the methods are described will allow scientists from all disciplines to see how these sometimes obscure-seeming methods can be applied to their own problems.

For those readers who work within my own profession I trust that the more 'generic' approach to the explanation and description of the techniques will still allow an understanding of how they may be applied to their own problems. There are, of course, some particular topics which only apply to molecular design and these have been included in Chapter 10 so for these readers I recommend the unusual approach of reading this book by starting at the end. The text also includes examples from the drug design field, in some cases very specific examples such as chemical library design, so I expect that this will be a useful handbook for the molecular designer.

1.1   INTRODUCTION

Most applications of data analysis involve attempts to fit a model, usually quantitative,[1] to a set of experimental measurements or observations. The reasons for fitting such models are varied. For example, the model may be purely empirical and be required in order to make predictions for new experiments. On the other hand, the model may be based on some theory or law, and an evaluation of the fit of the data to the model may be used to give insight into the processes underlying the observations made. In some cases the ability to fit a model to a set of data successfully may provide the inspiration to formulate some new hypothesis. The type of model which may be fitted to any set of data depends not only on the nature of the data (see Section 1.4) but also on the intended use of the model. In many applications a model is meant to be used predictively,

---

[1] According to the type of data involved, the model may be qualitative.

but the predictions need not necessarily be quantitative. Chapters 4 and 5 give examples of techniques which may be used to make qualitative predictions, as do the classification methods described in Chapter 7.

In some circumstances it may appear that data analysis is not fitting a model at all! The simple procedure of plotting the values of two variables against one another might not seem to be modelling, unless it is already known that the variables are related by some law (for example absorbance and concentration, related by Beer's law). The production of a bivariate plot may be thought of as fitting a model which is simply dictated by the variables. This may be an alien concept but it is a useful way of visualizing what is happening when multivariate techniques are used for the display of data (see Chapter 4). The resulting plots may be thought of as models which have been fitted by the data and as a result they give some insight into the information that the model, and hence the data, contains.

1.2 TYPES OF DATA

At this point it is necessary to introduce some jargon which will help to distinguish the two main types of data which are involved in data analysis. The observed or experimentally measured data which will be modelled is known as a *dependent variable* or variables if there are more than one. It is expected that this type of data will be determined by some features, properties or factors of the system under observation or experiment, and it will thus be dependent on (related by) some more or less complex function of these factors. It is often the aim of data analysis to predict values of one or more dependent variables from values of one or more *independent variables*. The independent variables are observed properties of the system under study which, although they may be dependent on other properties, are not dependent on the observed or experimental data of interest. I have tried to phrase this in the most general way to cover the largest number of applications but perhaps a few examples may serve to illustrate the point. Dependent variables are usually determined by experimental measurement or observation on some (hopefully) relevant test system. This may be a biological system such as a purified enzyme, cell culture, piece of tissue, or whole animal; alternatively it may be a panel of tasters, a measurement of viscosity, the brightness of a star, the size of a nanoparticle, the quantification of colour and so on. Independent variables may be determined experimentally, may be observed themselves, may be calculated or may be

ID	Response	Ind 1	Ind 2	Ind 3	Ind 4	Ind 5
Case 1	14	1.6	136	0.03	-12.6	19542
Case 2	24	2	197	0.07	-8.2	15005
Case 3	-6	9.05	211	0.1	-1	10098
Case 4	19	6	55	0.005	-0.99	17126
Case 5	88.2	3.66	126	0.8	0	19183
Case 6	43	12	83	0.79	-1.3	12087
....	....	....	....	....	....	....
....	....	....	....	....	....	....
Case n	11	7.05	156	0.05	-6.5	16345

Figure 1.1   Example of a dataset laid out as a table.

controlled by the investigator. Examples of independent variables are temperature, atmospheric pressure, time, molecular volume, concentration, distance, etc.

One other piece of jargon concerns the way that the elements of a data set are 'labelled'. The data set shown in Figure 1.1 is laid out as a table in the 'natural' way that most scientists would use; each row corresponds to a sample or experimental observation and each column corresponds to some measurement or observation (or calculation) for that row.

The rows are called 'cases' and they may correspond to a sample or an observation, say, at a time point, a compound that has been tested for its pharmacological activity, a food that has been treated in some way, a particular blend of materials and so on. The first column is a label, or case identifier, and subsequent columns are variables which may also be called descriptors or properties or features. In the example shown in the figure there is one case label, one dependent variable and five independent variables for n cases which may also be thought of as an n by 6 matrix (ignoring the case label column). This may be more generally written as an n by p matrix where p is the number of variables. There is nothing unusual in laying out a data set as a table. I expect most scientists did this for their first experiment, but the concept of thinking of a data set as a mathematical construct, a matrix, may not come so easily. Many of the techniques used for data analysis depend on matrix manipulations and although it isn't necessary to know the details of operations such as matrix multiplication in order to use them, thinking of a data set as a matrix does help to explain them.

Important features of data such as scales of measurement and distribution are described in later sections of this chapter but first we should consider the sources and nature of the data.

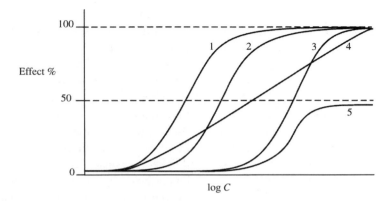

Figure 1.2 Typical and not so typical dose–response curves for a set of five different compounds.

1.3 SOURCES OF DATA

1.3.1 Dependent Data

Important considerations for dependent data are that their measurement should be well defined experimentally, and that they should be consistent amongst the cases (objects, samples, observations) in a set. This may seem obvious, and of course it is good scientific practice to ensure that an experiment is well controlled, but it is not always obvious that data is consistent, particularly when analysed by someone who did not generate it. Consider the set of curves shown in Figure 1.2 where biological effect is plotted against concentration.

Compounds 1–3 can be seen to be 'well behaved' in that their dose–response curves are of very similar shape and are just shifted along the concentration axis depending on their potency. Curves of this sigmoidal shape are quite typical; common practice is to take 50 % as the measure of effect and read off the concentration to achieve this from the dose axis. The advantage of this is that the curve is linear in this region; thus if the ED_{50} (the dose to give 50 % effect) has been bracketed by experimental measurements, it simply requires linear interpolation to obtain the ED_{50}. A further advantage of this procedure is that the effect is changing most rapidly with concentration in the 50 % part of the curve. Since small changes in concentration produce large changes in effect it is possible to get the most precise measure of the concentration required to cause a standard effect. The curve for compound 4 illustrates a common problem in that it does not run parallel to the others; this compound produces small effects (<50 %) at very low doses but needs comparatively high concentrations to achieve effects in excess of 50 %. Compound 5 demonstrates yet another deviation from the norm in that it does not achieve 50 % effect. There may be a variety of reasons for these deviations from the usual behaviour, such as changes in mechanism, solubility problems, and so on, but the effect is to produce inconsistent results which may be difficult or impossible to analyse.

The situation shown here where full dose–response data is available is very good from the point of view of the analyst, since it is relatively easy to detect abnormal behaviour and the data will have good precision. However, it is often time-consuming, expensive, or both, to collect such a full set of data. There is also the question of what is required from the test in terms of the eventual application. There is little point, for example, in making precise measurements in the millimolar range when the target activity must be of the order of micromolar or nanomolar. Thus, it should be borne in mind that the data available for analysis may not always be as good as it appears at first sight. Any time spent in a preliminary examination of the data and discussion with those involved in the measurement will usually be amply repaid.

1.3.2   Independent Data

Independent variables also should be well defined experimentally, or in terms of an observation or calculation protocol, and should also be consistent amongst the cases in a set. It is important to know the precision of the independent variables since they may be used to make predictions of a dependent variable. Obviously the precision, or lack of it, of the independent variables will control the precision of the predictions. Some data analysis techniques assume that all the error is in the dependent variable, which is rarely ever the case.

There are many different types of independent variables. Some may be controlled by an investigator as part of the experimental procedure. The length of time that something is heated, for example, and the temperature that it is heated to may be independent variables. Others may be obtained by observation or measurement but might not be under the control of the investigator. Consider the case of the prediction of tropical storms where measurements may be made over a period of time of ocean temperature, air pressure, relative humidity, wind speed and so on. Any or all of these parameters may be used as independent variables in attempts to model the development or duration of a tropical storm.

In the field of molecular design[2] the independent variables are most often physicochemical properties or molecular descriptors which characterize the molecules under study. There are a number of ways in which chemical structures can be characterized. Particular chemical features such as aromatic rings, carboxyl groups, chlorine atoms, double bonds and suchlike can be listed or counted. If they are listed, answering the question 'does the structure contain this feature?', then they will be binary descriptors taking the value of 1 for present and 0 for absent. If they are counts then the parameter will be a real valued number between 0 and some maximum value for the compounds in the set. Measured properties such as melting point, solubility, partition coefficient and so on are an obvious source of chemical descriptors. Other parameters, many of them, may be calculated from a knowledge of the 2-dimensional (2D) or 3-dimensional (3D) structure of the compounds [1, 2]. Actually, there are some descriptors, such as molecular weight, which don't even require a 2D structure.

1.4 THE NATURE OF DATA

One of the most frequently overlooked aspects of data analysis is consideration of the data that is going to be analysed. How accurate is it? How complete is it? How representative is it? These are some of the questions that should be asked about any set of data, preferably *before* starting to try and understand it, along with the general question 'what do the numbers, or symbols, or categories mean?'

So far, in this book the terms descriptor, parameter, and property have been used interchangeably. This can perhaps be justified in that it helps to avoid repetition, but they do actually mean different things and so it would be best to define them here. Descriptor refers to any means by which a sample (case, object) is described or characterized: for molecules the term aromatic, for example, is a descriptor, as are the quantities molecular weight and boiling point. Physicochemical property refers to a feature of a molecule which is determined by its physical or chemical properties, or a combination of both. Parameter is a term which is used

---

[2] Molecular design means the design of a biologically active substance such as a pharmaceutical or pesticide, or of a 'performance' chemical such as a fragrance, flavour, and so on or a formulation such as paint, adhesive, etc.

to refer to some numerical measure of a descriptor or physicochemical property. The two descriptors molecular weight and boiling point are also both parameters; the term aromatic is a descriptor but not a parameter, whereas the question 'How many aromatic rings?' gives rise to a parameter. All parameters are thus descriptors but not vice versa.

The next few sections discuss some of the more important aspects of the nature and properties of data. It is often the data itself that dictates which particular analytical method may be used to examine it and how successful the outcome of that examination will be.

1.4.1   Types of Data and Scales of Measurement

In the examples of descriptors and parameters given here it may have been noticed that there are differences in the 'nature' of the values used to express them. This is because variables, both dependent and independent, can be classified as *qualitative* or *quantitative*. Qualitative variables contain data that can be placed into distinct classes; 'dead' or 'alive', for example, 'hot' or 'cold', 'aromatic' or 'non-aromatic' are examples of binary or dichotomous qualitative variables. Quantitative variables contain data that is numerical and can be ranked or ordered. Examples of quantitative variables are length, temperature, age, weight, etc. Quantitative variables can be further divided into discrete or continuous. Discrete variables are usually counts such as 'how many objects in a group', 'number of hydroxyl groups', 'number of components in a mixture', and so on. Continuous variables, such as height, time, volume, etc. can assume any value within a given range.

In addition to the classification of variables as qualitative/quantitative and the further division into discrete/continuous, variables can also be classified according to how they are categorized, counted or measured. This is because of differences in the scales of measurement used for variables. It is necessary to consider four different scales of measurement: nominal, ordinal, interval, and ratio. It is important to be aware of the properties of these scales since the nature of the scales determines which analytical methods should be used to treat the data.

Nominal

This is the weakest level of measurement, i.e. has the lowest information content, and applies to the situation where a number or other symbol is used to assign membership to a class. The terms male and female, young and old, aromatic and non-aromatic are all descriptors based on nominal scales. These are dichotomous descriptors, in that the objects (people or compounds) belong to one class or another, but this is not the only type of nominal descriptor. Colour, subdivided into as many classes as desired, is a nominal descriptor as is the question 'which of the four halogens does the compound contain?'

Ordinal

Like the nominal scale, the ordinal scale of measurement places objects in different classes but here the classes bear some relation to one another, expressed by the term greater than (>). Thus, from the previous example, old > middle-aged > young. Two examples in the context of molecular design are toxic > slightly toxic > nontoxic, and fully saturated > partially saturated > unsaturated. The latter descriptor might also be represented by the number of double bonds present in the structures although this is not chemically equivalent since triple bonds are ignored. It is important to be aware of the situations in which a parameter might appear to be measured on an interval or ratio scale (see below), but because of the distribution of compounds in the set under study, these effectively become nominal or ordinal descriptors (see next section).

Interval

An interval scale has the characteristics of a nominal scale, but in addition the distances between any two numbers on the scale are of known size. The zero point and the units of measurement of an interval scale are arbitrary: a good example of an interval scale parameter is boiling point. This could be measured on either the Fahrenheit or Celsius temperature scales but the information content of the boiling point values is the same.

Ratio

A ratio scale is an interval scale which has a true zero point as its origin. Mass is an example of a parameter measured on a ratio scale, as are parameters which describe dimensions such as length, volume, etc. An additional property of the ratio scale, hinted at in the name, is that it contains a true ratio between values. A measurement of 200 for one sample and 100 for another, for example, means a ratio of 2:1 between these two samples.

What is the significance of these different scales of measurement? As will be discussed later, many of the well-known statistical methods are parametric, that is, they rely on assumptions concerning the distribution of the data. The computation of parametric tests involves arithmetic manipulation such as addition, multiplication, and division, and this should only be carried out on data measured on interval or ratio scales. When these procedures are used on data measured on other scales they introduce distortions into the data and thus cast doubt on any conclusions which may be drawn from the tests. Nonparametric or 'distribution-free' methods, on the other hand, concentrate on an order or ranking of data and thus can be used with ordinal data. Some of the nonparametric techniques are also designed to operate with classified (nominal) data. Since interval and ratio scales of measurement have all the properties of ordinal scales it is possible to use nonparametric methods for data measured on these scales. Thus, the distribution-free techniques are the 'safest' to use since they can be applied to most types of data. If, however, the data does conform to the distributional assumptions of the parametric techniques, these methods may well extract more information from the data.

1.4.2   Data Distribution

Statistics is often concerned with the treatment of a small[3] number of samples which have been drawn from a much larger population. Each of these samples may be described by one or more variables which have been measured or calculated for that sample. For each variable there exists a population of samples. It is the properties of these populations of variables that allows the assignment of probabilities, for example, the likelihood that the value of a variable will fall into a particular range, and the assessment of significance (i.e. is one number significantly different from another). Probability theory and statistics are, in fact, separate subjects; each may be said to be the inverse of the other, but for the purposes of this discussion they may be regarded as doing the same job.

---

[3] The term 'small' here may represent hundreds or even thousands of samples. This is a small number compared to a population which is often taken to be infinite.

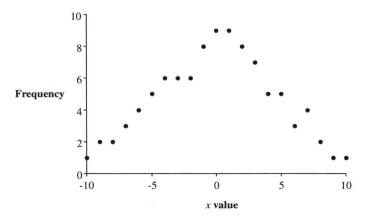

Figure 1.3 Frequency distribution for the variable x over the range −10 to +10.

How are the properties of the population used? Perhaps one of the most familiar concepts in statistics is the frequency distribution. A plot of a frequency distribution is shown in Figure 1.3, where the ordinate (y-axis) represents the number of occurrences of a particular value of a variable given by the scales of the abscissa (x-axis).

If the data is discrete, usually but not necessarily measured on nominal or ordinal scales, then the variable values can only correspond to the points marked on the scale on the abscissa. If the data is continuous, a problem arises in the creation of a frequency distribution, since every value in the data set may be different and the resultant plot would be a very uninteresting straight line at $y = 1$. This may be overcome by taking ranges of the variable and counting the number of occurrences of values within each range. For the example shown in Figure 1.4 (where there are a total of 50 values in all), the ranges are 0–1, 1–2, 2–3, and so on up to 9–10.

It can be seen that these points fall on a roughly bell-shaped curve with the largest number of occurrences of the variable occurring around the peak of the curve, corresponding to the mean of the set. The mean of the sample is given the symbol $\overline{X}$ and is obtained by summing all the sample values together and dividing by the number of samples as shown in Equation (1.1).

$$\overline{X} = \frac{x_1 + x_2 + x_3 + \ldots \ldots x_n}{n} = \frac{\sum x}{n} \tag{1.1}$$

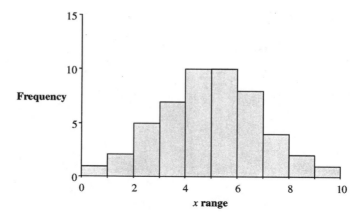

Figure 1.4 Frequency histogram for the continuous variable x over the range 0 to +10.

The mean, since it is derived from a sample, is known as a *statistic*. The corresponding value for a population, the population mean, is given the symbol μ and this is known as a *parameter*, another use for the term. A convention in statistics is that Greek letters are used to denote parameters (measures or characteristics of the population) and Roman letters are used for statistics. The mean is known as a 'measure of central tendency' (others are the mode, median and midrange) which means that it gives some idea of the centre of the distribution of the values of the variable. In addition to knowing the centre of the distribution it is important to know how the data values are spread through the distribution. Are they clustered around the mean or do they spread evenly throughout the distribution? Measures of distribution are often known as 'measures of dispersion' and the most often used are variance and standard deviation. Variance is the average of the squares of the distance of each data value from the mean as shown in Equation (1.2):

$$s^2 = \frac{\sum (X - \overline{X})^2}{n - 1} \tag{1.2}$$

The symbol used for the sample variance is s^2 which at first sight might appear strange. Why use the square sign in a symbol for a quantity like this? The reason is that the standard deviation (s) of a sample is the square root of the variance. The standard deviation has the same units as the units of the original variable whereas the variance has units that are the square of the original units. Another odd thing might be noticed

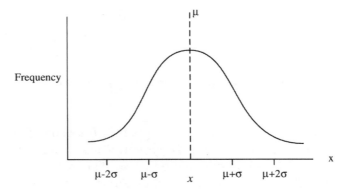

Figure 1.5   Probability distribution for a very large number of values of the variable x; μ equals the mean of the set and σ the standard deviation.

about Equation (1.2) and that is the use of $n - 1$ in the denominator. When calculating the mean the summation (Equation (1.1)) is divided by the number of data points, n, so why is $n - 1$ used here? The reason for this, apparently, is that the variance computed using n usually underestimates the population variance and thus the summation is divided by $n - 1$ giving a slightly larger value. The corresponding symbols for the population parameters are σ^2 for the variance and σ for the standard deviation. A graphical illustration of the meaning of μ and σ is shown in Figure 1.5, which is a frequency distribution like Figures 1.3 and 1.4 but with more data values so that we obtain a smooth curve.

The figure shows that μ is located in the centre of the distribution, as expected, and that the values of the variable x along the abscissa have been replaced by the mean +/− multiples of the standard deviation. This is because there is a theorem (Chebyshev's) which specifies the proportions of the spread of values in terms of the standard deviation, there is more on this later.

It is at this point that we can see a link between statistics and probability theory. If the height of the curve is standardized so that the area underneath it is unity, the graph is called a probability curve. The height of the curve at some point x can be denoted by $f(x)$ which is called the probability density function (p.d.f.). This function is such that it satisfies the condition that the area under the curve is unity

$$\int_{-\infty}^{\infty} f(x)\mathrm{d}x = 1 \qquad (1.3)$$

This now allows us to find the probability that a value of x will fall in any given range by finding the integral of the p.d.f. over that range:

$$probability \ (x_1 < x < x_2) = \int_{x_1}^{x_2} f(x)dx \qquad (1.4)$$

This brief and rather incomplete description of frequency distributions and their relationship to probability distribution has been for the purpose of introducing the normal distribution curve. The normal or Gaussian distribution is the most important of the distributions that are considered in statistics. The height of a normal distribution curve is given by

$$f(x) = \frac{1}{\sigma\sqrt{2\pi}}e^{-(x-\mu)^2/2\sigma^2} \qquad (1.5)$$

This rather complicated function was chosen so that the total area under the curve is equal to 1 for all values of μ and σ. Equation (1.5) has been given so that the connection between probability and the two parameters μ and σ of the distribution can be seen. The curve is shown in Figure 1.5 where the abscissa is marked in units of σ. It can be seen that the curve is symmetric about μ, the mean, which is a measure of the *location* or 'central tendency' of the distribution. As mentioned earlier, there is a theorem that specifies the proportion of the spread of values in any distribution. In the special case of the normal distribution this means that approximately 68 % of the data values will fall within 1 standard deviation of the mean and 95 % within 2 standard deviations. Put another way, about one observation in three will lie more than one standard deviation (σ) from the mean and about one observation in 20 will lie more than two standard deviations from the mean. The standard deviation is a measure of the *spread* or 'dispersion'; it is these two properties, location and spread, of a distribution which allow us to make estimates of likelihood (or 'significance').

Some other features of the normal distribution can be seen by consideration of Figure 1.6. In part (a) of the figure, the distribution is no longer symmetrical; there are more values of the variable with a higher value.

This distribution is said to be skewed, it has a positive skewness; the distribution shown in part (b) is said to be negatively skewed. In part (c) three distributions are overlaid which have differing degrees of 'steepness' of the curve around the mean. The statistical term used

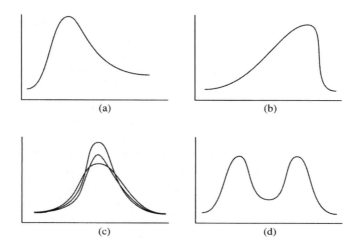

Figure 1.6 Illustration of deviations of probability distributions from a normal distribution.

to describe the steepness, or degree of peakedness, of a distribution is *kurtosis*. Various measures may be used to express kurtosis; one known as the *moment ratio* gives a value of three for a normal distribution. Thus it is possible to judge how far a distribution deviates from normality by calculating values of skewness (= 0 for a normal distribution) and kurtosis. As will be seen later, these measures of how 'well behaved' a variable is may be used as an aid to variable selection. Finally, in part (d) of Figure 1.6 it can be seen that the distribution appears to have two means. This is known as a *bimodal* distribution, which has its own particular set of properties distinct to those of the normal distribution.

1.4.3   Deviations in Distribution

There are many situations in which a variable that might be expected to have a normal distribution does not. Take for example the molecular weight of a set of assorted painkillers. If the compounds in the set consisted of aspirin and morphine derivatives, then we might see a bimodal distribution with two peaks corresponding to values of around 180 (mol.wt. of aspirin) and 285 (mol.wt. of morphine). Skewed and kurtosed distributions may arise for a variety of reasons, and the effect they will have on an analysis depends on the assumptions employed in the analysis and the degree to which the distributions deviate from normality, or whatever distribution is assumed. This, of course, is not a very satisfactory statement to someone who is asking the question, 'Is my data good enough (sufficiently well behaved) to apply such and such a method to it?' Unfortunately, there is not usually a simple answer to this sort of question. In general, the further the data deviates from the type of distribution that is assumed when a model is fitted, the less reliable will be the conclusions drawn from that model. It is worth pointing out here that real data is unlikely to conform perfectly to a normal distribution, or any other 'standard' distribution for that matter. Checking the distribution is necessary so that we know what type of method can be used to treat the data, as described later, and how reliable any estimates will be which are based on assumptions of distribution. A caution should be sounded here in that it is easy to become too critical and use a poor or less than 'perfect' distribution as an excuse not to use a particular technique, or to discount the results of an analysis.

Another problem which is frequently encountered in the distribution of data is the presence of outliers. Consider the data shown in Table 1.1 where calculated values of electrophilic superdelocalizability (ESDL10) are given for a set of analogues of antimycin A_1, compounds which kill human parasitic worms, *Dipetalonema vitae*.

The mean and standard deviation of this variable give no clues as to how well it is distributed and the skewness and kurtosis values of -3.15

Table 1.1 Physicochemical properties and antifilarial activity of antimycin analogues (reproduced from ref. [3] with permission from American Chemical Society).

Compound number	ESDL10	Calculated log P	Melting point °C	Activity
1	−0.3896	7.239	81	−0.845
2	−0.4706	5.960	183	−0.380
3	−0.4688	6.994	207	1.398
4	−0.4129	7.372	143	0.319
5	−0.3762	5.730	165	−0.875
6	−0.3280	6.994	192	0.824
7	−0.3649	6.755	256	1.839
8	−0.5404	6.695	199	1.020
9	−0.4499	7.372	151	0.420
10	−0.3473	5.670	195	0.000
11	−0.7942	4.888	212	0.097
12	−0.4057	6.205	246	1.130
13	−0.4094	6.113	208	0.920
14	−1.4855	6.180	159	0.770
15	−0.3427	5.681	178	0.301
16	−0.4597	6.838	222	1.357

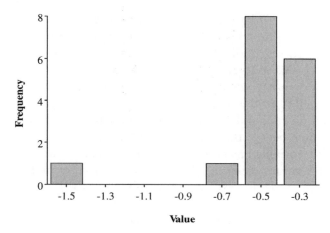

Figure 1.7 Frequency distribution for the variable ESDL10 given in Table 1.1.

and 10.65 respectively might not suggest that it deviates too seriously from normal. A frequency distribution for this variable, however, reveals the presence of a single extreme value (compound 14) as shown in Figure 1.7.

This data was analysed by multiple linear regression (discussed further in Chapter 6), which is a method based on properties of the normal distribution. The presence of this outlier had quite profound effects on the analysis, which could have been avoided if the data distribution had been checked at the outset (particularly by the present author). Outliers can be very informative and should not simply be discarded as so frequently happens. If an outlier is found in one of the descriptor variables (physicochemical data), then it may show that a mistake has been made in the measurement or calculation of that variable for that compound. In the case of properties derived from computational chemistry calculations it may indicate that some basic assumption has been violated or that the particular method employed was not appropriate for that compound. An example of this can be found in semi-empirical molecular orbital methods which are only parameterized for a limited set of the elements. Outliers are not always due to mistakes, however. Consider the calculation of electrostatic potential around a molecule. It is easy to identify regions of high and low values, and these are often used to provide criteria for alignment or as a pictorial explanation of biological properties. The value of an electrostatic potential minimum or maximum, or the value of the potential at a given point, has been used as a parameter to describe sets of molecules. This is fine as long as each molecule in the set has a maximum and/or minimum at approximately the same place. Problems arise if a small number of the structures do not have the corresponding values in which case they will form outliers. The effect of this is to cause the variable, apparently measured on an interval scale, to become a nominal descriptor. Take, for example, the case where 80 % of the members of the set have an electrostatic potential minimum of around −50 kcal/mole at a particular position. For the remaining members of the set, the electrostatic potential at this position is zero. This variable has now become an 'indicator' variable which has two distinct values (zero for 20 % of the molecules and −50 for the remainder) that identify two different subsets of the data. The problem may be overcome if the magnitude of a minimum or maximum is taken, irrespective of position, although problems may occur with molecules that have multiple minima or maxima. There is also the more difficult philosophical question of what do such values 'mean'.

When outliers occur in the biological or dependent data, they may also indicate mistakes: perhaps the wrong compound was tested, or it did not dissolve, a result was misrecorded, or the test did not work out as expected. However, in dependent data sets, outliers may be even more informative. They may indicate a change in biological mechanism, or perhaps they demonstrate that some important structural feature has been altered or a critical value of a physicochemical property exceeded. Once again, it is best not to simply discard such outliers, they may be very informative.

Is there anything that can be done to improve a poorly distributed variable? The answer is yes, but it is a qualified yes since the use of too many 'tricks' to improve distribution may introduce other distortions which will obscure useful patterns in the data. The first step in improving distribution is to identify outliers and then, if possible, identify the cause(s) of such outliers. If an outlier cannot be 'fixed' it may need to be removed from the data set. The second step involves the consideration of the rest of the values in the set. If a variable has a high value of kurtosis or skewness, is there some good reason for this? Does the variable really measure what we think it does? Are the calculations/measurements sound for all of the members of the set, particularly at the extremes of the range for skewed distributions or around the mean where kurtosis is a problem. Finally, would a transformation help? Taking the logarithm of a variable will often make it behave more like a normally distributed variable, but this is not a justification for always taking logs!

A final point on the matter of data distribution concerns the non-parametric methods. Although these techniques are not based on distributional assumptions, they may still suffer from the effects of 'strange' distributions in the data. The presence of outliers or the effective conversion of interval to ordinal data, as in the electrostatic potential example, may lead to misleading results.

1.5   ANALYTICAL METHODS

This whole book is concerned with analytical methods, as the following chapters will show, so the purpose of this section is to introduce and explain some of the terms which are used to describe the techniques. These terms, like most jargon, also often serve to obscure the methodology to the casual or novice user so it is hoped that this section will help to unveil the techniques.

First, we should consider some of the expressions which are used to describe the methods in general. *Biometrics* is a term which has been used since the early 20th century to describe the development of mathematical and statistical methods to data analysis problems in the biological sciences. *Chemometrics* is used to describe 'any mathematical or statistical procedure which is used to analyse chemical data' [4]. Thus, the simple act of plotting a calibration curve is chemometrics, as is the process of fitting a line to that plot by the method of least squares, as is the analysis by principal components of the spectrum of a solution containing several species. Any chemist who carries out quantitative experiments is also a chemometrician! *Univariate statistics* is (perhaps unsurprisingly) the term given to describe the statistical analysis of a single variable. This is the type of statistics which is normally taught on an introductory course; it involves the analysis of variance of a single variable to give quantities such as the mean and standard deviation, and some measures of the distribution of the data. *Multivariate statistics* describes the application of statistical methods to more than one variable at a time, and is perhaps more useful than univariate methods since most problems in real life are multivariate. We might more correctly use the term *multivariate analysis* since not all multivariate methods are statistical. Chemometrics and multivariate analysis refer to more or less the same things, chemometrics being the broader term since it includes univariate techniques.[4]

Pattern recognition is the name given to any method which helps to reveal the patterns within a data set. A definition of pattern recognition is that it 'seeks similarities and regularities present in the data'. Some

---

[4] But, of course, it is restricted to chemical problems.

Table 1.2 Anaesthetic activity and hydrophobicity of a series of alcohols (reproduced from ref. [5] with permission from American Society for Pharmacology and Experimental Therapeutics (ASPET)).

Alcohol	$\Sigma\pi$	Anaesthetic activity $(\log 1/C)$
C_2H_5OH	1.0	0.481
$n{-}C_3H_7OH$	1.5	0.959
$n{-}C_4H_9OH$	2.0	1.523
$n{-}C_5H_{11}OH$	2.5	2.152
$n{-}C_7H_{15}OH$	3.5	3.420
$n{-}C_8H_{17}OH$	4.0	3.886
$n{-}C_9H_{19}OH$	4.5	4.602
$n{-}C_{10}H_{21}OH$	5.0	5.00
$n{-}C_{11}H_{23}OH$	5.5	5.301
$n{-}C_{12}H_{25}OH$	6.0	5.124

of the display techniques described in Chapter 4 are quite obvious examples of pattern recognition since they result in a visual display of the patterns in data. However, consider the data shown in Table 1.2 where the anaesthetic activity of a series of alcohols is given as the logarithm of the reciprocal of the concentration needed to induce a particular level of anaesthesia.

The other column in this table $(\Sigma\pi)$ is a measure of the hydrophobicity of each of the alcohols. Hydrophobicity, which means literally 'water hating', reflects the tendency of molecules to partition into membranes in a biological system (see Chapter 10 for more detail) and is a physicochemical descriptor of the alcohols. Inspection of the table reveals a fairly obvious relationship between log 1/C and $\Sigma\pi$ but this is most easily seen by a plot as shown in Figure 1.8.

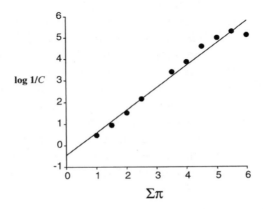

Figure 1.8 Plot of biological response (log 1/C) against $\Sigma\pi$ (from Table 1.2).

This relationship can be expressed in a very concise form as shown in Equation (1.6):

$$\log {}^1\!/_C = 1.039 \sum \pi - 0.442 \qquad (1.6)$$

This is an example of a simple linear regression equation. Regression equations and the statistics which may be used to describe their 'goodness of fit', to a linear or other model, are explained in detail in Chapter 6. For the purposes of demonstrating this relationship it is sufficient to say that the values of the logarithm of a reciprocal concentration ($\log 1/C$) in Equation (1.6) are obtained by multiplication of the $\Sigma\pi$ values by a coefficient (1.039) and the addition of a constant term (-0.442). The equation is shown in graphical form (Figure 1.8); the slope of the fitted line is equal to the regression coefficient (1.039) and the intercept of the line with the zero point of the x-axis is equal to the constant (-0.442). Thus, the pattern obvious in the data table may be shown by the simple bivariate plot and expressed numerically in Equation (1.6). These are examples of pattern recognition although regression models would not normally be classed as pattern recognition methods.

Pattern recognition and chemometrics are more or less synonymous. Some of the pattern recognition techniques are derived from research into artificial intelligence. We can 'borrow' some useful jargon from this field which is related to the concept of 'training' an algorithm or device to carry out a particular task. Suppose that we have a set of data which describes a collection of compounds which can be classified as active or inactive in some biological test. The descriptor data, or independent variables, may be whole molecule parameters such as melting point, or may be substituent constants, or may be calculated quantities such as molecular orbital energies. One simple way in which this data may be analysed is to compare the values of the variables for the active compounds with those of the inactives (see discriminant analysis in Chapter 7). This may enable one to establish a rule or rules which will distinguish the two classes. For example, all the actives may have melting points above $250\,^\circ$C and/or may have highest occupied molecular orbital (HOMO) energy values below -10.5. The production of these rules, by inspection of the data or by use of an algorithm, is called *supervised learning* since knowledge of class membership was used to generate them. The dependent variable, in this case membership of the active or inactive class, is used in the learning or training process. *Unsupervised learning*, on the other hand, does not make use of a dependent variable.

An example of unsupervised learning for this data set might be to plot the values of two of the descriptor variables against one another. Class membership for the compounds could then be marked on the plot and a pattern may be seen to emerge from the data. If we chose melting point and HOMO as the two variables to plot, we may see a grouping of the active compounds where HOMO < -10.5 and melting point $>250\,^{\circ}$C.

The distinction between supervised and unsupervised learning may seem unimportant but there is a significant philosophical difference between the two. When we seek a rule to classify data, there is a possibility that any apparent rule may happen by chance. It may, for example, be a coincidence that all the active compounds have high melting points; in such a case the rule will not be predictive. This may be misleading, embarrassing, expensive, or all three! Chance effects may also occur with unsupervised learning but are much less likely since unsupervised learning does not seek to generate rules. Chance effects are discussed in more detail in Chapters 6 and 7. The concept of learning may also be used to define some data sets. A set of compounds which have already been tested in some biological system, or which are about to be tested, is known as a *learning* or *training set*. In the case of a supervised learning method this data will be used to train the technique but this term applies equally well to the unsupervised case. Judicious choice of the training set will have profound effects on the success of the application of any analytical method, supervised or unsupervised, since the information contained in this set dictates the information that can be extracted (see Chapter 2). A set of untested or yet to be synthesized compounds is called a *test set*, the objective of data analysis usually being to make predictions for the test set (also sometimes called a *prediction set*). A further type of data set, known as an *evaluation set*, may also be used. This consists of a set of compounds for which test results are available but which is not used in the construction of the model. Examination of the prediction results for an evaluation set can give some insight into the validity and accuracy of the model.

Finally we should define the terms *parametric* and *nonparametric*. A measure of the distribution of a variable (see Section 1.4.2) is a measure of one of the parameters of that variable. If we had measurements for all possible values of a variable (an infinite number of measurements), then we would be able to compute a value for the population distribution. Statistics is concerned with a much smaller set of measurements which forms a sample of that population and for which we can calculate a sample distribution. A well-known example of this is the Gaussian or normal distribution. One of the assumptions made in statistics is that a sample distribution, which we can measure, will behave like a population distribution which we cannot. Although population distributions cannot be measured, some of their properties can be predicted by theory. Many statistical methods are based on the properties of population distributions, particularly the normal distribution. These are called *parametric techniques* since they make use of the distribution parameter. Before using a parametric method, the distribution of the variables involved should be calculated. This is very often ignored, although fortunately many of the techniques based on assumptions about the normal distribution are quite robust to departures from normality. There are also techniques which do not rely on the properties of a distribution, and these are known as *nonparametric* or '*distribution free*' methods.

1.6 SUMMARY

In this chapter the following points were covered:

1. dependent and independent variables and how data tables are laid out;
2. where data comes from and some of its properties;
3. descriptors, parameters and properties; .
4. nominal, ordinal, interval and ratio scales;
5. frequency distributions, the normal distribution, definition and explanation of mean, variance and standard deviation. skewness and kurtosis;
6. the difference between sample and population properties;
7. factors causing deviations in distribution;
8. terminology – univariate and multivariate statistics, chemometrics and biometrics, pattern recognition, supervised and unsupervised learning. Training, test and evaluation sets, parametric and nonparametric or 'distribution free' techniques.

REFERENCES

[1] Livingstone, D.J. (2000). 'The Characterization of Chemical Structures Using Molecular Properties. A Survey', *Journal of Chemical Information and Computer Science*, **40**, 195–209.
[2] Livingstone, D.J. (2003). 'Theoretical Property Predictions', *Current Topics in Medicinal Chemistry*, **3**, 1171–92.

[3] Selwood, D.L, Livingstone, D.J., Comley, J.C.W. *et al.* (1990). *Journal of Medicinal Chemistry*, **33**, 136–42.

[4] Kowalski, B., Brown, S., and Van de Ginste, B. (1987). *Journal of Chemometrics*, **1**, 1–2.

[5] Hansch, C., Steward, A. R., Iwasa, J., and Deutsch, E.W. (1965). *Molecular Pharmacology*, **1**, 205–13.

2

Experimental Design: Experiment and Set Selection

Points covered in this chapter

- What is experimental design?
- Experimental design terminology
- Experimental design techniques
- Compound and set selection
- High throughput experiments

2.1 WHAT IS EXPERIMENTAL DESIGN?

All experiments are designed insofar as decisions are made concerning the choice of apparatus, reagents, animals, analytical instruments, temperature, solvent, and so on. Such decisions need to be made for any individual experiment, or series of experiments, and will be based on prior experience, reference to the literature, or perhaps the whim of an individual experimentalist. How can we be sure that we have made the right decisions? Does it matter whether we have made the right decisions? After all, it can be argued that an experiment is just that; the results obtained with a particular experimental set-up are the results obtained, and as such are of more or less interest depending on what they are. To some extent the reason for conducting the experiment in the first place may decide whether the question of the right decisions matters. If the experiment is being carried out to comply with some legislation from

A Practical Guide to Scientific Data Analysis   David Livingstone
© 2009 John Wiley & Sons, Ltd

a regulatory body (for example, toxicity testing for a new drug may require administration at doses which are fixed multiples of the therapeutic dose), then the experimental decisions do not matter. Alternatively the experiment may be intended to synthesize a new compound. In this case, if the target compound is produced then all is well, except that we do not know that the yield obtained is the best we could get by that route. This may not matter if we are just interested in having a sample of the compound, but what should we do if the experiment does not produce the compound? The experiment can be repeated using different conditions: for example, we could change the temperature or the time taken for a particular step, the solvent, or solvent mixture, and perhaps the reagents. These experimental variables are called factors and even quite a simple chemical synthesis may involve a number of factors.

What is the best way to set about altering these factors to achieve the desired goal of synthesizing the compound? We could try 'trial and error', and indeed many people do, but this is unlikely to be the most efficient way of investigating the effect of these factors, unless we are lucky. However, the most important feature of experimental design lies in the difference between 'population' values and 'sample' values. As described in the last chapter, any experimental result, whether a measurement or the yield from a synthesis, comes from a population of such results. When we do an experiment we wish to know about the population structure (values) using a sample to give some idea of population behaviour. In general, the larger the number of samples obtained, the better our idea of population values. The advantages of well-designed experiments are that the information can be obtained with minimum sample sizes and that the results can be interpreted to give the population information required. The next section gives some examples of strategies for experimental design. This can be of use directly in the planning of experiments but will also introduce some concepts which are of considerable importance in the analysis of property–activity data for drug design.

One may ask the question 'how is experimental design relevant to the analysis of biological data when the experimental determinations have already been made?'. One of the factors which is important in the testing of a set of compounds, and indeed intended to be the most important, is the nature of the compounds used. This set of compounds is called a training set, and selection of an appropriate training set will help to ensure that the optimum information is extracted from the experimental measurements made on the set. As will be shown in Section 2.3, the choice of training set may also determine the most appropriate physicochemical descriptors to use in the analysis of experimental data for the set.

At the risk of stating the obvious, it should be pointed out that the application of any analytical method to a training set can only extract as much information as the set contains. Careful selection of the training set can help to ensure that the information it contains is maximized.

2.2 EXPERIMENTAL DESIGN TECHNIQUES

Before discussing the techniques of experimental design it is necessary to introduce some terms which describe the important features of experiments. As mentioned in the previous section, the variables which determine the outcome of an experiment are called *factors*. Factors may be qualitative or quantitative. As an example, consider an experiment which is intended to assess how well a compound or set of compounds acts as an inhibitor of an enzyme *in vitro*. The enzyme assay will be carried out at a certain temperature and pH using a particular buffer with a given substrate and perhaps cofactor at fixed concentrations. Different buffers may be employed, as might different substrates if the enzyme catalyses a class of reaction (e.g. angiotensin converting enzyme splits off dipeptides from the C-terminal end of peptides with widely varying terminal amino acid sequences). These are qualitative factors since to change them is an 'all or nothing' change. The other factors such as temperature, pH, and the concentration of reagents are quantitative; for quantitative factors it is necessary to decide the levels which they can adopt. Most enzymes carry out their catalytic function best at a particular pH and temperature, and will cease to function at all if the conditions are changed too far from this optimum. In the case of human enzymes, for example, the optimum temperature is likely to be 37 °C and the range of temperature over which they catalyse reactions may be (say) 32–42 °C. Thus we may choose three levels for this factor: low, medium, and high, corresponding to 32, 37, and 42 °C. The reason for choosing a small number of discrete levels for a continuous variable such as this is to reduce the number of possible experiments (as will be seen below). In the case of an enzyme assay, experience might lead us to expect that medium would give the highest turnover of substrate although experimental convenience might prompt the use of a different level of this factor.[1]

---

[1] Optimum enzyme performance might deplete the substrate too rapidly and give an inaccurate measure of a compound as an inhibitor.

Table 2.1 Experimental factors for an enzyme assay.

Factor	Type	Levels	Treatments
Temperature	Quantitative	32 °C, 37 °C, 42 °C	3
Cofactor	Qualitative	Yes/No	2
pH	Quantitative	7.0, 7.8	2
		Total =	12

A particular set of experimental conditions is known as a *treatment* and for any experiment there are as many possible treatments as the product of the levels of each of the factors involved. Suppose that we wish to investigate the performance of an enzyme with respect to temperature, pH, and the presence or absence of a natural cofactor. The substrate concentration might be fixed at its physiological level and we might choose two levels of pH which we expect to bracket the optimum pH. Here the cofactor is a qualitative factor which can adopt one of two levels, present or absent, temperature may take three levels as before, and pH has two levels, thus there are $2 \times 3 \times 2 = 12$ possible treatments, as shown in Table 2.1. The outcome of an experiment for a given treatment is termed a *response*; in this enzyme example the response might be the rate of conversion of substrate, and in our previous example the response might be the percentage yield of compound synthesized.

How can we tell the importance of the effect of a given factor on a response and how can we tell if this apparent effect is real? For example, the effect may be a population property rather than a sample property due to random variation. This can be achieved by *replication*, the larger the number of *replicates* of a given treatment then the better will be our estimate of the variation in response for that treatment. We will also have greater confidence that any one result obtained is not spurious since we can compare it with the others and thus compare variation due to the treatment to random variation. Replication, however, consumes resources such as time and material, and so an important feature of experimental design is to *balance* the effort between replication and change in treatment. A balanced design is one in which the treatments to be compared are replicated the same number of times, and this is desirable because it maintains orthogonality between factors (an important assumption in the analysis of variance).

The factors which have been discussed so far are susceptible to change by the experimentalist and are thus referred to as controlled factors. Other factors may also affect the experimental response and these are referred to as uncontrolled factors. How can experiments be designed to detect, and hopefully eliminate, the effects of uncontrolled factors on the response? Uncontrolled factors may very often be time-dependent. In the example of the enzyme assay, the substrate concentration may be monitored using an instrument such as a spectrophotometer. The response of the instrument may change with time and this might be confused with effects due to the different treatments unless steps are taken to avoid this. One approach might be to calibrate the instrument at regular intervals with a standard solution: calibration is, of course, a routine procedure. However, this approach might fail if the standard solution were subject to change with time, unless fresh solutions were made for each calibration. Even if the more obvious time-dependent uncontrolled factors such as instrument drift are accounted for, there may be other important factors at work.

One way to help eliminate the effect of uncontrolled factors is to randomize the order in which the different treatments are applied. The consideration that the order in which experiments are carried out is important introduces the concept of batches, known as *blocks*, of experiments. Since an individual experiment takes a certain amount of time and will require a given amount of material it may not be possible to carry out all of the required treatments on the same day or with the same batch of reagents. If the enzyme assay takes one hour to complete, it may not be possible to examine more than six treatments in a day. Taking just the factor pH and considering three levels, low (7.2), medium (7.4), and high (7.6), labelled as A, B, and C, a randomized block design with two replicates might be

A, B, B, C, A, C

Another assay might take less time and allow eight treatments to be carried out in one day. This block of experiments would enable us to examine the effect of two factors at two levels with two replicates. Taking the factors pH and cofactor, labelled as A and B for high and low levels of pH, and 1 and 0 for presence or absence of cofactor, a randomized block design with two replicates might be

A1, B0, B1, A0, A1, B0, B1, A0

This has the advantage that the presence or absence of cofactor alternates between treatments but has the disadvantage that the high pH treatments with cofactor occur at the beginning and in the middle of the block. If an instrument is switched on at the beginning of the day and then again half-way through the day, say after a lunch break, then the replicates of this particular treatment will be subject to a unique set of conditions – the one-hour warm-up period of the instrument. Similarly the low pH treatments are carried out at the same times after the instrument is switched on. This particular set of eight treatments might be better split into two blocks of four; in order to keep blocks of experiments homogeneous it pays to keep them as small as possible. Alternatively, better randomization within the block of eight treatments would help to guard against uncontrolled factors. Once again, balance is important, it may be better to examine the effect of one factor in a block of experiments using a larger number of replicates. This is the way that block designs are usually employed, examining the effect of one experimental factor while holding other factors constant. This does introduce the added complication of possible differences between the blocks. In a blocked design, the effect of a factor is of interest, not normally the effect of the blocks, so the solution is to ensure good randomization within the block and/or to repeat the block of experimental treatments.

A summary of the terms introduced so far is shown in Table 2.2

Table 2.2   Terms used in experimental design.

Term	Meaning
Response	The outcome of an experiment
Factor	A variable which affects the experimental response. These can be controlled and uncontrolled, qualitative and quantitative
Level	The values which a factor can adopt. In the case of a qualitative factor these are usually binary (e.g. present/absent)
Treatment	Conditions for a given experiment, e.g. temperature, pH, reagent concentration, solvent
Block	Set of experimental treatments carried out in a particular time-period or with a particular batch of material and thus (hopefully) under the same conditions. Generally, observations within a block can be compared with greater precision than between blocks
Randomization	Random ordering of the treatments within a block in an attempt to minimize the effect of uncontrolled factors
Replication	Repeat experimental treatments to estimate the significance of the effect of individual factors on the response (and to identify 'unusual' effects)
Balance	The relationship between the number of treatments to be compared and the number of replicates of each treatment examined

2.2.1 Single-factor Design Methods

The block design shown previously is referred to as a 'balanced, complete block design' since all of the treatments were examined within the block (hence 'complete'), and the number of replicates was the same for all treatments (hence 'balanced'). If the number of treatments and their replicates is larger than the number of experimental 'slots' in a block then it will be necessary to carry out two or more blocks of experiments to examine the effect of the factor. This requires that the blocks of experiments are chosen in such a way that comparisons between treatments will not be affected. When all of the comparisons are of equal importance (for example, low vs. high temperature, low vs. medium, and high vs. medium) the treatments should be selected in a balanced way so that any two occur together the same number of times as any other two. This type of experimental design is known as 'balanced, incomplete block design'. The results of this type of design are more difficult to analyse than the results of a complete design, but easier than if the treatments were chosen at random which would be an 'unbalanced, incomplete block design'.

The time taken for an individual experiment may determine how many experiments can be carried out in a block, as may the amount of material required for each treatment. If both of these factors, or any other two 'blocking variables', are important then it is necessary to organize the treatments to take account of two (potential) uncontrolled factors. Suppose that: there are three possible treatments, A, B, and C; it is only possible to examine three treatments in a day; a given batch of material is sufficient for three treatments; time of day is considered to be an important factor. A randomized design for this is shown below.

Batch	Time of day		
1	A	B	C
2	B	C	A
3	C	A	B

This is known as a *Latin square*, perhaps the best-known term in experimental design, and is used to ensure that the treatments are randomized to avoid trends within the design. Thus, the Latin square design is used when considering the effect of one factor and two blocking variables. In this case the factor was divided into three levels giving rise to three treatments: this requires a 3 × 3 matrix. If the factor has more levels, then the design will simply be a larger symmetrical matrix, i.e. 4 × 4, 5 × 5, and so on. What about the situation where there are three blocking variables? In the enzyme assay example, time of day may be important and there may only be sufficient cofactor in one batch for three assays and similarly only sufficient enzyme in one batch for three assays. This calls for a design known as a *Graeco-Latin square* which is made by superimposing two different Latin squares. There are two possible 3 × 3 Latin squares:

A	B	C
B	C	A
C	A	B

A	B	C
C	A	B
B	C	A

The 3 × 3 Graeco-Latin square is made by the superimposition of these two Latin squares with the third blocking variable denoted by Greek letters thus:

Aα	Bβ	Cγ
Bγ	Cα	Aβ
Cβ	Aγ	Bα

It can be seen in this design that each treatment occurs only once in each row and column (two of the blocking variables, say time of day and cofactor batch) and only once with each level (α, β, and γ) of the third blocking variable, the enzyme batch. Both Latin squares and Graeco-Latin squares (and *Hyper-Graeco-Latin squares* for more blocking variables) are most effective if they are replicated and are also subject to the rules of randomization which apply to simple block designs. While these designs are useful in situations where only one experimental factor is varied, it is clear that if several factors are important (a more usual situation), this approach will require a large number of experiments to examine their effects. Another disadvantage of designing experiments to investigate a single factor at a time is that the interactions between factors are not examined since in this approach all other factors are kept constant.

Table 2.3   Options in a multiple-factor design.

Solvent	Temperature	
	T_1	T_2
S_1	y_1	y_3
S_2	y_2	y_4

2.2.2   Factorial Design (Multiple-factor Design)

The simplest example of the consideration of multiple experimental factors would involve two factors. Taking the earlier example of a chemical synthesis, suppose that we were interested in the effect of two different reaction temperatures, T_1 and T_2, and two different solvents, S_1 and S_2, on the yield of the reaction. The minimum number of experiments required to give us information on both factors is three, one at T_1S_1 (y_1), a second at T_1S_2 (y_2) involving change in solvent, and a third at T_2S_1 (y_3) involving a change in temperature (see Table 2.3). The effect of changing temperature is given by the difference in yields $y_3 - y_1$ and the effect of changing solvent is given by $y_2 - y_1$. Confirmation of these results could be obtained by duplication of the above requiring a total of six experiments.

This is a 'one variable at a time' approach since each factor is examined separately. However, if a fourth experiment, T_2S_2 (y_4), is added to Table 2.3 we now have two measures of the effect of changing each factor but only require four experiments. In addition to saving two experimental determinations, this approach allows the detection of interaction effects between the factors, such as the effect of changing temperature in solvent 2 ($y_4 - y_2$) compared with solvent 1 ($y_3 - y_1$). The factorial approach is not only more efficient in terms of the number of experiments required and the identification of interaction effects, it can also be useful in optimization. For example, having estimated the main effects and interaction terms of some experimental factors it may be possible to predict the likely combinations of these factors which will give an optimum response. One drawback to this procedure is that it may not always be possible to establish all possible combinations of treatments, resulting in an unbalanced design. Factorial designs also tend to involve a large number of experiments, the investigation of three factors at three levels, for example, requires 27 runs (3^f where f is the number of factors) without replication of any of the combinations. However, it is possible to reduce the number of experiments required as will be shown later.

A nice example of the use of factorial design in chemical synthesis was published by Coleman and co-workers [1]. The reaction of 1,1,1-trichloro-3-methyl-3-phospholene (1) with methanol produces 1-methoxy-3-methyl-2-phospholene oxide (2) as shown in the reaction scheme. The experimental procedure involved the slow addition of a known quantity of methanol to a known quantity of 1 in dichloromethane held at subambient temperature. The mixture was then stirred until it reached ambient temperature and neutralized with aqueous sodium carbonate solution; the product was extracted with dichloromethane.

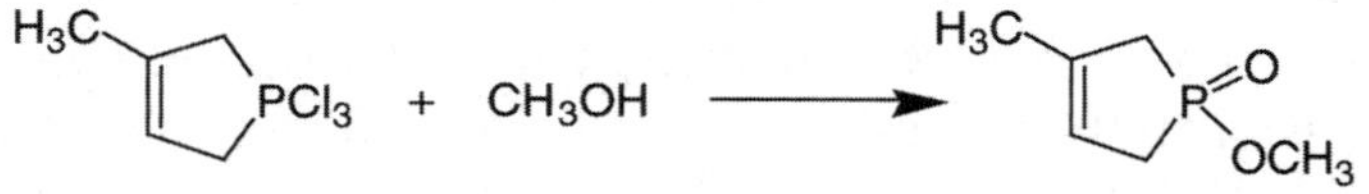

Scheme 2.1 .

The yield from this reaction was 25 % and could not significantly be improved by changing one variable (concentration, temperature, addition time, etc.) at a time. Three variables were chosen for investigation by factorial design using two levels of each.

A: Addition temperature (-15 or $0\ ^{\circ}$C)
B: Concentration of 1 (50 or 100 g in 400 cm^3 dichloromethane)
C: Addition time of methanol (one or four hours)

This led to eight different treatments (2^3), which resulted in several yields above 25 % (as shown in Table 2.4), the largest being 42.5 %.

The effect on an experimental response due to a factor is called a *main* effect whereas the effect caused by one factor at each level of the other factor is called an *interaction* effect (two way). The larger the number of levels of the factors studied in a factorial design, the higher the order of the interaction effects that can be identified. In a three-level factorial design it is possible to detect quadratic effects although it is often difficult to interpret the information. Three-level factorial designs also require a considerable number of experiments (3^f) as shown above. For this reason it is often found convenient to consider factors at just two levels, high/low or yes/no, to give 2^f factorial designs.

Table 2.4 Responses from full factorial design (reproduced from ref. [1] with permission of the Royal Society of Chemistry).

Order of treatment	Treatment combination[a]	Yield (%)
3	–	24.8
6	a	42.5
1	b	39.0
7	ab	18.2
2	c	32.8
4	ac	33.0
8	bc	13.2
5	abc	24.3

[a]Where a lower-case letter is shown, this indicates that a particular factor was used at its high level in that treatment, e.g. a means an addition temperature of 0 °C. When a letter is missing the factor was at its low level.

Another feature of these full factorial designs, full in the sense that all combinations of all levels of each factor are considered, is that interactions between multiple factors may be identified. In a factorial design with six factors at two levels ($2^6 = 64$ experiments) there are six main effects (for the six factors), 15 two-factor interactions (two-way effects), 20 three-factor, 15 four-factor, 6 five-factor, and 1 six-factor interactions. Are these interactions all likely to be important? The answer, fortunately, is no. In general, main effects tend to be larger than two-factor interactions which in turn tend to be larger than three-factor interactions and so on. Because these higher order interaction terms tend not to be significant it is possible to devise smaller factorial designs which will still investigate the experimental factor space efficiently but which will require far fewer experiments. It is also often found that in factorial designs with many experimental factors, only a few factors are important. These smaller factorial designs are referred to as *fractional factorial designs*, where the fraction is defined as the ratio of the number of experimental runs needed in a full design. For example, the full factorial design for five factors at two levels requires 32 (2^5) runs: if this is investigated in 16 experiments it is a half-fraction factorial design. Fractional designs may also be designated as 2^{f-n} where f is the number of factors as before and n is the number of half-fractions, 2^{5-1} is a half-fraction factorial design in five factors, 2^{6-2} is a quarter-fraction design in six factors.

Of course, it is rare in life to get something for nothing and that principle applies to fractional factorial designs. Although a fractional design allows one to investigate an experimental system with the expenditure of less effort, it is achieved at the expense of clarity in our ability to separate main effects from interactions. The response obtained from certain treatments could be caused by the main effect of one factor or a two-(three-, four-, five-, etc.) factor interaction. These effects are said to be confounded; because they are indistinguishable from one another, they are also said to be aliases of one another. It is the choice of aliases which lies at the heart of successful fractional factorial design. As mentioned before, we might expect that main effects would be more significant than two-factor effects which will be more important than three-factor effects. The aim of fractional design is thus to alias main effects and two-factor effects with as high-order interaction terms as possible.

The phospholene oxide synthesis mentioned earlier provides a good example of the use of fractional factorial design. Having carried out the full factorial design in three factors (addition temperature, concentration of phospholene, and addition time) further experiments were made to 'fine-tune' the response. These probing experiments involved small changes to one factor while the others were held constant in order to determine whether an optimum had been reached in the synthetic conditions. Figure 2.1 shows a response surface for the high addition time in which percentage yield is plotted against phospholene concentration and addition temperature. The response surface is quite complex and demonstrates that a maximum yield had not been achieved for the factors examined in the first full factorial design. In fact the largest yield found in these probing experiments was 57 %, a reasonable increase over the highest yield of 42.5 % shown in Table 2.4.

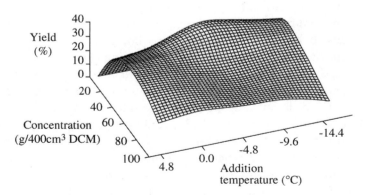

Figure 2.1   Response surface for phospholene oxide synthesis (reproduced from ref. [1] with permission of the Royal Society of Chemistry).

Table 2.5 Responses from fractional factorial design (reproduced from ref. [1] with permission of the Royal Society of Chemistry).

Treatment combination[a]	Yield	Aliasing effect
–	45.1	
ad	60.2	A with BD
bde	62.5	B with CE +AD
abe	46.8	D with AB
ce	77.8	C with BE
acde	49.8	AC with DE
bcd	53.6	E with BC
abc	70.8	AE with CD

[a] As explained in Table 2.4.

The shape of the response surface suggests the involvement of other factors in the yield of this reaction and three more experimental variables were identified: concentration of methanol, stirring time, and temperature. Fixing the concentration of phospholene at 25 g in 400 cm³ of dichloromethane (a broad peak on the response surface) leaves five experimental factors to consider, requiring a total of 32 (2^5) experiments to investigate them. These experiments were split into four blocks of eight and hence each block is a quarter-fraction of 32 experiments. The results for the first block are shown in Table 2.5, the experimental factors being

A: Addition temperature (-10 or $0°$ C)
B: Addition time of methanol (15 or 30 minutes)
C: Concentration of methanol (136 or 272 cm³)
D: Stirring time (0.5 or 2 hours)
E: Stirring temperature (addition temperature or ambient)

This particular block of eight runs was generated by aliasing D with AB and also E with BC, after carrying out full 2^3 experiments of A, B, and C. As can be seen from Table 2.5, the best yield from this principal block of experiments, which contains variables and variable interactions expected to be important, was 78 %, a considerable improvement over the previously found best yield of 57 %. Having identified important factors, or combinations of factors with which they are aliased, it is possible to choose other treatment combinations which will clarify the situation. The best yield obtained for this synthesis was 90 % using treatment combination e (addition temperature -10 °C, addition time 15 mins, methanol concentration 136 cm³, stirring time 0.5 hours, stirring temperature ambient).

2.2.3   D-optimal Design

Factorial design methods offer the advantage of a systematic exploration of the factors that are likely to affect the outcome of an experiment; they also allow the identification of interactions between these factors. They suffer from the disadvantage that they may require a large number of experiments, particularly if several levels of each factor are to be examined. This can be overcome to some extent by the use of fractional factorial designs although the aliasing of multi-factor interactions with main effects can be a disadvantage. Perhaps one of the biggest disadvantages of factorial and fractional factorial designs in chemistry is the need to specify different levels for the factors. If a factorial design approach is to be successful, it is necessary to construct treatment combinations which explore the factor space. When the experimental factors are variables such as time, temperature, and concentration, this usually presents few problems. However, if the factors are related to chemical structure, such as the choice of compounds for testing, the situation may be quite different. A factorial design may require a particular compound which is very difficult to synthesize. Alternatively, a design may call for a particular set of physicochemical properties which cannot be achieved, such as a very small, very hydrophobic substituent.

The philosophy of the factorial approach is attractive, so, are there related techniques which are more appropriate to the special requirements of chemistry? There are a number of other methods for experimental design but one that is becoming applied in several chemical applications is known as 'D-optimal design'. The origin of the expression 'D-optimal' is a bit of statistical jargon based on the determinant of the variance–covariance matrix. As will be seen in the next section on compound selection, a well-chosen set of experiments (or compounds) will have a wide spread in the experimental factor space (variance). A well-chosen set will also be such that the correlation (see Box 2.1) between experimental factors is at a minimum (covariance). The determinant of a matrix is a single number and in the case of a variance–covariance matrix for a data set this number describes the 'balance' between variance and covariance. This determinant will be a maximum for experiments or compounds which have maximum variance and minimum covariance, and thus the optimization of the determinant (D-optimal) is the basis of the design. Examples of the use of D-optimal design are given in the next section and later in this book.

Box 2.1 The correlation coefficient, r

An important property of any variable, which is used in many statistical operations is a quantity called the variance, V. The variance is a measure of how the values of a variable are distributed about the mean and is defined by

$$V = \sum_{i=1}^{n}(x_i - \overline{x})^2/n$$

where $\overline{x}$ is the mean of the set of x values and the summation is carried out over all n members of the set. When values are available for two or more variables describing a set of objects (compounds, samples, etc.) a related quantity may be calculated called the covariance, $C_{(x,y)}$. The covariance is a measure of how the values of one variable (x) are distributed about their mean compared with how the corresponding values of another variable (y) are distributed about their mean. Covariance is defined as

$$C_{(x,y)} = \sum_{i=1}^{n}(x_i - \overline{x})(y_i - \overline{y})/n$$

The covariance has some useful properties. If the values of variable x change in the same way as the values of variable y, the covariance will be positive. For small values of x, $x - \overline{x}$ will be negative and $y - \overline{y}$ will be negative yielding a positive product. For large values of x, $x - \overline{x}$ will be positive as will $y - \overline{y}$ and thus the summation yields a positive number. If, on the other hand, y decreases as x increases the covariance will be negative. The sign of the covariance of the two variables indicates how they change with respect to one another: positive if they go up and down together, negative if one increases as the other decreases. But is it possible to say how clearly one variable mirrors the change in another? The answer is yes, by the calculation of a quantity known as the correlation coefficient

$$r = C_{(x,y)}/\left[V_{(x)} \times V_{(y)}\right]^{\frac{1}{2}}$$

Division of the covariance by the square root of the product of the individual variances allows us to put a scale on the degree to which two variables are related. If y changes by exactly the same amount as x changes, and in the same direction, the correlation coefficient is $+1$. If y decreases by exactly the same amount as x increases the correlation coefficient is -1. If the changes in y are completely unrelated to the changes in x, the correlation coefficient will be 0.

Further discussion of other experimental design techniques, with the exception of *simplex* optimization (see next section), is outside the scope of this book. Hopefully this section will have introduced the principles of experimental design; the reader interested in further details should consult one of the excellent texts available which deal with this subject in detail [2–5]. A review discusses the application of experimental design techniques to chemical synthesis [6].

2.3 STRATEGIES FOR COMPOUND SELECTION

This section shows the development of methods for the selection of compounds in computer-aided molecular design but the methods employed are equally relevant to the selection of data sets in many scientific applications. It could also have been entitled strategies for 'training set selection' where compounds are the members of a training set. Training sets are required whenever a new biological test is established, when compounds are selected from an archive for screening in an existing test, or when a set of biological (or other) data is to be analysed. It cannot be stressed sufficiently that selection of appropriate training sets is crucial to the success of new synthetic programmes, screening, and analysis. The following examples illustrate various aspects of compound selection.

Some of the earliest techniques for compound selection were essentially visual and as such have considerable appeal compared with the (apparently) more complex statistical and mathematical methods. The first method to be reported came from a study of the relationships between a set of commonly used substituent constants [7]. The stated purpose of this work was to examine the interdependence of these parameters and, as expected, correlations (see Box 2.1) were found between the hydrophobicity descriptor, π, and a number of 'bulk' parameters such as molecular volume and parachor (see Chapter 10 for an explanation of these descriptors). Why should interdependence between substituent constants be important? There are a number of answers to this question, as discussed further in this book, but for the present it is sufficient to say that independence between parameters is required so that clearer, perhaps mechanistic, conclusions might be drawn from correlations. As part of the investigation Craig plotted various parameters together, for example the plot of σ vs. π shown in Figure 2.2; such plots have since become known as Craig plots.

This diagram nicely illustrates the concept of a physicochemical parameter space. If we regard these two properties as potentially important

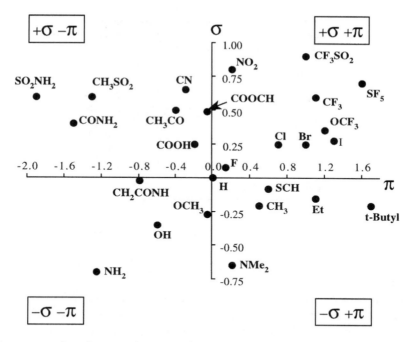

Figure 2.2 Plot of σ vs π for a set of common substituents (reproduced from ref. [7] with permission of the American Chemical Society).

experimental factors, in the sense that they are likely to control or at least influence experiments carried out using the compounds, then we should seek to choose substituents that span the parameter space. This is equivalent to the choice of experimental treatments which are intended to span the space of the experimental factors.

It is easy to see how substituents may be selected from a plot such as that shown in Figure 2.2, but will this ensure that the series is well chosen? The answer is no for two reasons. First, the choice of compounds based on the parameter space defined by just two substituent constants ignores the potential importance of any other factors. What is required is the selection of points in a multidimensional space, where each dimension corresponds to a physicochemical parameter, so that the space is sampled evenly. This is described later in this section. The second problem with compound choice based on sampling a two-parameter space concerns the correlation between the parameters. Table 2.6 lists a set of substituents with their corresponding π values.

At first sight this might appear to be a well-chosen set since the substituents cover a range of -1.2 to $+1.4$ log units in π and are represented

Table 2.6 π values for a set of substituents (reproduced from ref. [8] copyright 1984 with permission from Elsevier).

Substituent	π
NH_2	−1.23
OH	−0.67
OCH_3	−0.02
H	0.00
F	0.14
Cl	0.70
Br	0.86
SCF_3	1.44

at fairly even steps over the range. If, however, we now list the σ values for these substituents, as shown in Table 2.7, we see that they also span a good range of σ but that the two sets of values correspond to one another.

In general, there is no correlation between π and σ as can be seen from the scatter of points in Figure 2.2. For this particular set of substituents, however, there is a high correlation of 0.95; in trying to rationalize the biological properties of this set it would not be possible to distinguish between electronic and hydrophobic effects. There are other consequences of such correlations between parameters, known as collinearity, which involve multiple regression (Chapter 6), data display (Chapter 4), and other multivariate methods (Chapters 7 and 8). This is discussed in the next chapter and in the chapters which detail the techniques.

So, the two main problems in compound selection are the choice of analogues to sample effectively a multi-parameter space and the avoidance of collinearity between physicochemical descriptors. A number of

Table 2.7 π and σ values for the substituents in Table 2.6 (reproduced from ref. [8] copyright 1984 with permission from Elsevier).

Substituent	π	σ
NH_2	−1.23	−0.66
OH	−0.67	−0.37
OCH_3	−0.02	−0.27
H	0.00	0.00
F	0.14	0.06
Cl	0.70	0.23
Br	0.86	0.23
SCF_3	1.44	0.50

methods have been proposed to deal with these two problems. An attractive approach was published by Hansch and co-workers [9] which made use of cluster analysis (Chapter 5) to group 90 substituents described by five physicochemical parameters. Briefly, cluster analysis operates by the use of measurements of the distances between pairs of objects in multidimensional space using a distance such as the familiar Euclidean distance. Objects (compounds) which are close together in space become members of a single cluster. For a given level of similarity (i.e. value of the distance measure) a given number of clusters will be formed for a particular data set. At decreasing levels of similarity (greater values of the distance measure) further objects or clusters will be joined to the original clusters until eventually all objects in the set belong to a single cluster. The results of cluster analysis are most often reported in the form of a diagram known as a dendrogram (Figure 2.3).

A given level of similarity on the dendrogram gives rise to a particular number of clusters and thus it was possible for Hansch and his co-workers to produce lists of substituents belonging to 5, 10, 20, and 60 cluster sets. This allows a medicinal chemist to choose a substituent from each cluster when making a particular number of training set compounds (5, 10, 20, or 60) to help ensure that parameter space is well spanned. This work was subsequently updated to cover 166 substituents described by six parameters; lists of the cluster members were reported in the substituent constant book [11], sadly no longer in print although an

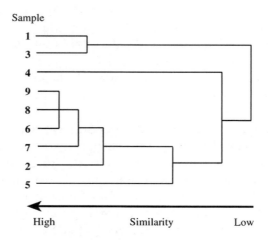

Figure 2.3 Example of a similarity diagram (dendrogram) from cluster analysis (reprinted from ref. [10] with permission of Elsevier).

Table 2.8 Examples of substituents belonging to clusters in the ten cluster set (reproduced from ref. [11] with permission of John Wiley & Sons, Inc.).

Cluster set number	Number of members	Examples of substituents
1	26	$-Br$, $-Cl$, $-NNN$, $-CH_3$, $-CH_2Br$
2	17	$-SO_2F$, $-NO_2$, $-CN$, $-1-Tetrazolyl$, $-SOCH_3$
3	2	$-IO_2$,, $-N(CH_3)_3$
4	8	$-OH$, $-NH_2$, $-NHCH_3$, $-NHC_4H_9$, $-NHC_6H_5$
5	18	$-CH_2OH$, $-NHCN$, $-NHCOCH_3$, $-CO_2H$, $-CONH_2$
6	21	$-OCF_3$, $-CH_2CN$, $-SCN$, $-CO_2CH_3$, $-CHO$
7	25	$-NCS$, $-Pyrryl$, $-OCOC_3H_7$, $-COC_6H_5$, $-OC_6H_5$
8	20	$-CH_2I$, $-C_6H_5$, $-C_5H_{11}$, $-Cyclohexyl$, $-C_4H_9$
9	21	$-NHC=S(NH_2)$, $-CONHC_3H_7$, $-NHCOC_2H_5$, $-C(OH)(CF_3)_2$, $-NHSO_2C_6H_5$
10	8	$-OC_4H_9$, $-N(CH_3)_2$, $-N(C_2H_5)_2$

updated listing has been published [12]. Table 2.8 lists some of the substituent members of the ten cluster set.

Another approach which makes use of the distances between points in a multidimensional space was published by Wootton and co-workers [13]. In this method the distances between each pair of substituents is calculated, as described for cluster analysis, and substituents are chosen in a stepwise fashion such that they exceed a certain preset minimum distance. The procedure requires the choice of a particular starting compound, probably but not necessarily the unsubstituted parent, and choice of the minimum distance. Figure 2.4 gives an illustration of this process to the choice of eight substituents from a set of 35. The resulting correlation between the two parameters for this set was low (-0.05). A related technique has been described by Franke [14] in which principal components (see Chapter 4) are calculated from the physicochemical descriptor data and interpoint distances are calculated based on the principal components. Several techniques are compared in the reference cited.

These techniques for compound selection have relied on the choice of substituents such that the physicochemical parameter space is well covered; the resulting sets of compounds tend to be well spread and interparameter correlations low. These were the two criteria set out earlier for successful compound choice, although other criteria, such as synthetic feasibility, may be considered important [15]. An alternative way to deal with the problem of compound selection is to treat the physicochemical properties as experimental factors and apply the

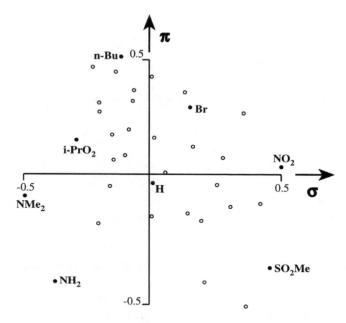

Figure 2.4 Example of the choice of substituents by multidimensional mapping (reproduced from ref. [13] with permission of the American Chemical Society).

techniques of factorial design. As described in Section 2.2, it is necessary to decide how many levels need to be considered for each individual factor in order to determine how many experimental treatments are required. Since the number of experiments (and hence compounds) increases as the product of the factor levels, it is usual to consider just two levels, say high and low, for each factor. This also allows qualitative factors such as the presence/absence of some functional group or structural feature to be included in the design. Of course, if some particular property is known or suspected to be of importance, then this may be considered at more than two levels. A major advantage of factorial design is that many factors may be considered at once and that interactions between factors may be identified, unlike the two parameter treatment of Craig plots. A disadvantage of factorial design is the large number of experiments that may need to be considered, but this may be reduced by the use of fractional factorials as described in Section 2.2. Austel [16] was the first to describe factorial designs for compound selection and he demonstrated the utility of this approach by application to literature

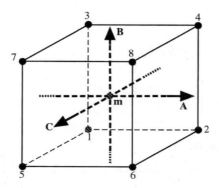

Figure 2.5 Representation of a factorial design in three factors (A, B, and C) (reproduced from ref. [16] copyright Elsevier (1982)).

examples. The relationship of a full to a half-fractional design is nicely illustrated in Figure 2.5.

The cube represents the space defined by three physicochemical properties A, B, and C and the points at the vertices represent the compounds chosen to examine various combinations of these parameters as shown in Table 2.9.

An extra point can usefully be considered in designs such as this corresponding to the midpoint of the factor space. If A, B, and C are substituent constants such as π, σ, and MR which are scaled to H = 0, this midpoint is the unsubstituted parent. A fractional factorial design in these three parameters is shown in Table 2.10. This fractional design investigates the main effects of parameters A and B, factor C is confounded (aliased, see Section 2.2.2) with interaction of A and B.

Table 2.9 Factorial design for three parameters (two levels) (reproduced from ref. [16] copyright Elsevier (1982)).

Compound	A	B	C
1	−	−	−
2	+	−	−
3	−	+	−
4	+	+	−
5	−	−	+
6	+	−	+
7	−	+	+
8	+	+	+

Table 2.10 Fractional factorial design for three parameters (two levels) (reproduced from ref. [16] copyright Elsevier (1982)).

Compound	A	B	C
1(5[a])	−	−	+
2(2)	+	−	−
3(3)	−	+	−
4(8)	+	+	+

[a]( ) corresponding compound in Table 2.9.

The four compounds in this table correspond to compounds 2, 3, 5, and 8 from Table 2.9 and in Figure 2.5 form the vertices of a regular tetrahedron, thus providing a good exploration of the three-dimensional factor space.

The investigation of the biological properties of peptide analogues gives a particularly striking illustration of the usefulness of fractional factorial design in the choice of analogues to examine. The problem with peptides is that any single amino acid may be replaced by any of the 20 coded amino acids, to say nothing of amino acid analogues. If a peptide of interest is varied in just four positions, it is possible to synthesize 160 000 (20^4) analogues. As pointed out by Hellberg *et al.* [17] who applied fractional factorial design to four series of peptides, the development of automated peptide synthesis has removed the problem of *how* to make peptide analogues. The major problem is *which* analogues to make. In order to apply the principles of experimental design to this problem it is necessary to define experimental factors (physicochemical properties) to be explored. These workers used 'principal properties' which were derived from the application of principal component analysis (see Chapter 4) to a data matrix of 29 physicochemical variables which describe the amino acids. The principal component analysis gave three new variables, labelled Z_1, Z_2, and Z_3, which were interpreted as being related to hydrophobicity (partition), bulk, and electronic properties respectively. Table 2.11 lists the values of these descriptor scales.

This is very similar to an earlier treatment of physicochemical properties by Cramer [18, 19], the so-called $BC(DEF)$ scales. The Z descriptor scales thus represent a three-dimensional property space for the amino acids. If only two levels are considered for each descriptor, high (+) and low (−), a full factorial design for substitution at one amino acid position in a peptide would require eight analogues. While this is a saving compared with the 20 possible analogues that could be made, a full

Table 2.11   Descriptor scales for the 20 'natural' amino acids (reproduced from ref. [17] with permission from American Chemical Society).

Acid	Z_1	Z_2	Z_3
Ala	0.07	−1.73	0.09
Val	−2.69	−2.53	−1.29
Leu	−4.19	−1.03	−0.98
Ile	−4.44	−1.68	−1.03
Pro	−1.22	0.88	2.23
Phe	−4.92	1.30	0.45
Trp	−4.75	3.65	0.85
Met	−2.49	−0.27	−0.41
Lys	2.84	1.41	−3.14
Arg	2.88	2.52	−3.44
His	2.41	1.74	1.11
Gly	2.23	−5.36	0.30
Ser	1.96	−1.63	0.57
Thr	0.92	−2.09	−1.40
Cys	0.71	−0.97	4.13
Tyr	−1.39	2.32	0.01
Asn	3.22	1.45	0.84
Gln	2.18	0.53	−1.14
Asp	3.64	1.13	2.36
Glu	3.08	0.39	−0.07

factorial design is impractical when multiple substitution positions are considered. A full design for four amino acid positions requires 4096 (8^4) analogues, for example. Hellberg suggested 1/2 (for one position), 1/8th and smaller fractional designs as shown in Table 2.12.

One of the problems with the use of factorial methods for compound selection is that it may be difficult or impossible to obtain a compound required for a particular treatment combination, either because the

Table 2.12   Number of peptide analogues required for fractional factorial design based on three Z scales (reproduced from ref. [17] with permission from American Chemical Society).

Number of varied positions	Minimum number of analogues
1	4
2	8
3–5	16
6–10	32
11–21	64

Table 2.13 Subsets of ten substituents from 35 chosen by four different methods (reprinted from ref. [21] with permission from Elsevier).

Method[a]			
h	**D**	**V(1)**	**V(2)**
H	H	H	Me
n-butyl	n-butyl	phenyl	t-butyl
phenyl	t-butyl	OH	OEt
CF_3	O–phenyl	O–phenyl	O–n-amyl
O–phenyl	NH_2	NMe_2	NH_2
NH_2	NMe_2	NO_2	NMe_2
NMe_2	NO_2	COOEt	NO_2
NO_2	SO_2Me	$CONH_2$	$COO(Me)_2$
SO_2NH_2	SO_2NH_2	SO_2Me	SO_2NH_2
F	F	Br	F
V = 1.698	1.750	1.437	1.356
D = 1.038	1.041	0.487	0.449
h = 1.614	1.589	1.502	1.420

[a]The methods are: h – maximization of information content, D – D-optimal design, V(1) maximal variance [13] and V(2) maximal variance [12].

synthesis is difficult or because that particular set of factors does not exist. One way to overcome this problem, as discussed in Section 2.2.3, is D-optimal design. Unger [20] has reported the application of a D-optimal design procedure to the selection of substituents from a set of 171, described by seven parameters.[2] The determinant of the variance–covariance matrix for the selected set of 20 substituents was 3.35×10^{11} which was 100 times better than the largest value (2.73×10^9) obtained in 100 simulations in which 20 substituents were randomly chosen. Herrmann [21] has compared the use of D-optimal design, two variance maximization methods, and an information-content maximization technique for compound selection. The results of the application of these strategies to the selection of ten substituents from a set of 35 are shown in Table 2.13. Both the D-optimal and the information-content methods produced better sets of substituents, as measured by variance (V) or determinant (D) values, than the variance maximization techniques.

The final compound selection procedures which will be mentioned here are the sequential simplex and the 'Topliss tree'. The sequential

[2] The reference includes listings of programs (written in APL) for compound selection by D-optimal design.

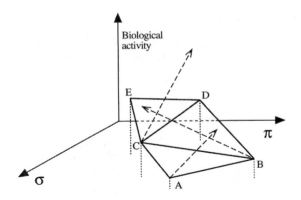

Figure 2.6 Illustration of the sequential simplex process of compound selection (reproduced from ref. [22] with permission of the American Chemical Society).

simplex, first reported by Darvas [22], is an application of a well-known optimization method which can be carried out graphically. Figure 2.6 shows three compounds, A, B, and C, plotted in a two-dimensional property space, say π and σ but any two properties may be used. Biological results are obtained for the three compounds and they are ranked in order of activity.

These compounds form a triangle in the two-dimensional property space, a new compound is chosen by the construction of a new triangle. The two most active compounds, say B and C for this example, form two of the vertices of the new triangle and the third vertex is found by taking a point opposite to the least active (A) to give the new triangle BCD. The new compound is tested and the activities of the three compounds compared, if B is now the least active then a new triangle CDE is constructed as shown in the figure. The procedure can be repeated until no further improvement in activity is obtained, or until all of the attainable physicochemical property space has been explored. This method is attractive in its simplicity and the fact that it requires no more complicated equipment than a piece of graph paper. The procedure is designed to ensure that an optimum is found in the particular parameters chosen so its success as a compound selection method is dependent on the correct choice of physicochemical properties. One of the problems with this method is that a compound may not exist that corresponds to a required simplex point. The simplex procedure is intended to operate with continuous experimental variables such as temperature, pressure, concentration, etc. There are other problems with the simplex procedure, for example, it requires biological activity data, but it has a number of advantages, not least of which being that *any* selection procedure is better than none.

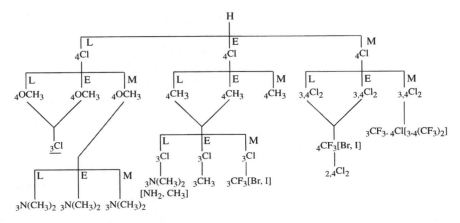

Figure 2.7 Illustration of the 'Topliss tree' process for compound selection; L, E, and M represent less, equal, and more active respectively (reproduced from ref. [23] with permission of Elsevier).

The 'Topliss tree' is an operational scheme that is designed to explore a given physicochemical property space in an efficient manner and is thus related to the sequential simplex approach [23]. In the case of aromatic substitution, for example, this approach assumes that the unsubstituted compound and the *para*-chloro derivative have both been made and tested. The activity of these two compounds are compared and the next substituent is chosen according to whether the chloro substituent displays higher, lower, or equal activity. This is shown schematically in Figure 2.7.

The rationale for the choice of $-OCH_3$ or Cl as the next substituent is based on the supposition that the given effects are dependent on changes in π or σ and, to a lesser extent, steric effects. This decision tree and the analogous scheme for aliphatic substitution are useful in that they suggest a systematic way in which compounds should be chosen. It suffers, perhaps, from the fact that it needs to start from a particular point, the unsubstituted compound, and that it requires data to guide it. Other schemes starting with different substituents could of course be drawn up and, like the simplex, any selection scheme is better than none.

2.4 HIGH THROUGHPUT EXPERIMENTS

A revolution happened in the pharmaceutical industry in the 1990s that was to have far-reaching consequences. The revolution, called 'combinatorial chemistry', took place in the medicinal chemistry departments of the research divisions. Hitherto, new compounds for screening as potential new drugs had been synthesized on an individual basis usually based on some design concept such as similarity to existing successful compounds, predictions from a mathematical model or expected interaction with a biological molecule based on molecular modelling. The ideas behind combinatorial chemistry were inspired by the existence of machines which could automatically synthesize polypeptides (small proteins) and latterly sequence them.

These processes were modified so that nonpeptides could be automatically synthesized since, although peptides often show remarkable biological properties, they are poor candidates for new drugs since there are problems with delivery and stability in the body. We routinely destroy proteins as fuel and the body is well prepared to identify and eliminate 'foreign' molecules such as peptides and proteins. Combinatorial chemistry strategies began with the synthesis of mixtures of compounds, at first a few tens or hundreds but then progressing to millions, but soon developed into parallel synthesis which is capable of producing very large numbers of single compounds. The two approaches, mixtures and singles, are both used today to produce libraries or arrays of compounds suitable for testing. But, what about experimental design? At first it was thought that the production and subsequent testing of such large numbers of compounds was bound to produce the required results in a suitable timeframe and thus design would be unnecessary. A little contemplation of the numbers involved, however, soon suggested that this would not be the case. As an example, decapeptides (a very small peptide of 10 residues) built from the 20 naturally occurring amino acids would have 20^{10} different sequences, in other words 1.024×10^{13} different molecules [24]. An even more startling example is given by the relatively small protein chymotrypsinogen-B which is composed of 245 amino acid residues. There are 20^{245} possible sequences -5.65×10^{318} molecules. An estimate of the number of particles in the visible universe is 10^{88} so there isn't enough available to build even a single molecule of every possible sequence! [25] The result of the realization of the enormous potential of combinatorial chemistry soon led to the development of design strategies for 'diverse' and 'focussed' libraries which, as the names imply, are intended to explore molecular diversity or to home in on a particularly promising range of chemical structures. There is also the question of the size of libraries. At first it was thought that large libraries were best, since this would maximize the chances of finding a useful compound, but it soon became evident that there were associated costs with combinatorial libraries, i.e. the cost of screening, and thus what was required was a library or screening collection that was large 'enough' but not too large [26].

So, what was the effect of this change in compound production on pharmaceutical research? As may be imagined, the effect was quite dramatic. Biological tests (screens) which were able to handle, at most, a few tens of compounds a day were quite inadequate for the examination of these new compound collections. With great ingenuity screening procedures were automated and designed to operate with small volumes of reagents so as to minimize the costs of testing. The new testing procedures became know as High Throughput Screening (HTS) and even Ultra-HTS. The laboratory instrument suppliers responded with greater automation of sample handling and specialist companies sprang up to supply robotic systems for liquid and solid sample handling. A typical website (http://www.htscreening.net/home/) provides numerous examples in this field.

The 'far-reaching' consequences of combinatorial chemistry and HTS are now spreading beyond the pharmaceutical industry. The agrochemicals industry, which faces similar challenges to pharmaceuticals, was an obvious early adopter and other specialist chemicals businesses are now following suit. Other materials such as catalysts can also benefit from these approaches and academic institutions are now beginning to pursue this approach (e.g. http://www.hts.ku.edu/). The whole process of combinatorial sample production and automated HTS is likely to be an important part of scientific research for the foreseeable future.

2.5 SUMMARY

The concepts underlying experimental design are to a great extent 'common sense' although the means to implement them may not be quite so obvious. The value of design, whether applied to an individual experiment or to the construction of a training set, should be clear from the examples shown in this chapter. Failure to apply some sort of design strategy may lead to a set of results which contain suboptimal information, at best, or which contain no useful information at all, at worst. Various design procedures may be applied to individual experiments, as indicated in the previous sections, and there are specialist reports which deal with topics such as synthesis [6]. A detailed review of design strategies which may be applied to the selection of compounds has been reported by Pleiss and Unger [27]. The development of combinatorial chemistry and its effect on compound screening to produce HTS methods has had a dramatic effect on pharmaceutical and agrochemical research which is now finding its way into many other industries and, eventually, many other areas of scientific research.

In this chapter the following points were covered:

1. the value of experimental design;
2. experimental factors – controlled and uncontrolled;
3. replication, randomization, experimental blocks and balance in design;
4. Latin squares, factorial and fractional factorial design;
5. main effects, interaction effects and the aliasing of effects;
6. variance, covariance and correlation;
7. the balance between maximization of variance and minimization of covariance;
8. D-optimal design and the sequential simplex for compound selection;
9. Combinatorial chemistry and high throughput screening.

REFERENCES

[1] Coleman, G.V., Price, D., Horrocks, A.R., and Stephenson, J.E. (1993). *Journal of the Chemical Society-Perkin Transactions II*, 629–32.
[2] Box, G.E.P., Hunter, W.J., and Hunter, J.S. (1978). *Statistics for experimentalists*. John Wiley & Sons, Inc, New York.
[3] Morgan, E. (1991). *Chemometrics: experimental design*. John Wiley & Sons, Ltd, Chichester.
[4] Montgomery, D.C. (2000). *Design and Analysis of Experiments*, 5th Edition. John Wiley & Sons, Ltd, Chichester.
[5] Ruxton, G.D. and Colegrave, N. (2003). *Experimental Design for the Life Sciences*. Oxford University Press, Oxford.
[6] Carlson, R. and Nordahl, A. (1993). *Topics in Current Chemistry*, **166**, 1–64.
[7] Craig, P.N. (1971). *Journal of Medicinal Chemistry*, **14**, 680–4.
[8] Franke, R. (1984). *Theoretical Drug Design Methods*. Vol 7 of *Pharmacochemistry library* (ed. W. Th. Nauta and R.F. Rekker), p. 145. Elsevier, Amsterdam.
[9] Hansch, C., Unger, S.H., and Forsythe, A.B. (1973). *Journal of Medicinal Chemistry*, **16**, 1217–22.
[10] Reibnegger, G., Werner-Felmayer, G., and Wachter, H. (1993). *Journal of Molecular Graphics*, **11**, 129–33.
[11] Hansch, C. and Leo, A.J. (1979). *Substituent constants for correlation analysis in chemistry and biology*, p. 58, John Wiley & Sons, Inc., New York.
[12] Hansch, C. and Leo, A.J. (1995). *Exploring QSAR. Fundamentals and Applications in Chemistry and Biology*, American Chemical Society, Washington, DC.
[13] Wootton, R., Cranfield, R., Sheppey, G.C., and Goodford, P.J. (1975). *Journal of Medicinal Chemistry*, **18**, 607–13.

[14] Streich, W.J., Dove, S., and Franke, R. (1980). *Journal of Medicinal Chemistry*, **23**, 1452–6.
[15] Schaper, K.-J. (1983). *Quantitative Structure—Activity Relationships*, **2**, 111–20.
[16] Austel, V. (1982). *European Journal of Medicinal Chemistry*, **17**, 9–16.
[17] Hellberg, S., Sjostrom, M., Skagerberg, B., and Wold, S. (1987). *Journal of Medicinal Chemistry*, **30**, 1126–35.
[18] Cramer, R.D. (1980). *Journal of the American Chemical Society*, **102**, 1837–49.
[19] Cramer, R.D. (1980). *Journal of the American Chemical Society*, **102**, 1849–59.
[20] Unger, S.H. (1980). In *Drug design*, (ed. E. J. Ariens), Vol. **9**, pp. 47–119. Academic Press, London.
[21] Herrmann, E.C. (1983). In *Quantitative Approaches to Drug Design* (ed. J. C. Dearden), pp. 231–2. Elsevier, Amsterdam.
[22] Darvas, F. (1974). *Journal of Medicinal Chemistry*, **17**, 799–804.
[23] Topliss, J.G. and Martin, Y.C. (1975). In *Drug Design* (ed. E.J. Ariens), Vol. **5**, pp. 1–21. Academic Press, London.
[24] Eason, M.A.M. and Rees, D.C. (2002). In *Medicinal Chemistry. Principles and Practice*, (ed. F. King), pp. 359–81. Royal Society of Chemistry, Cambridge, UK.
[25] Reproduced with the kind permission of Professor Furka (http://szerves.chem. elte.hu/Furka/).
[26] Lipkin, M.J., Stevens, A.P., Livingstone, D.J. and Harris, C.J. (2008). *Combinatorial Chemistry & High Throughput Screening*, **11**, 482–93.
[27] Pleiss, M.A. and Unger, S.H. (1990). In *Quantitative Drug Design* (ed. C.A. Ramsden), Vol. 4 of *Comprehensive Medicinal Chemistry. The Rational Design, Mechanistic Study and Therapeutic Application of Chemical Compounds*, (ed. C. Hansch, P.G. Sammes, and J.B. Taylor), pp. 561–87. Pergamon Press, Oxford.

3

Data Pre-treatment and Variable Selection

Points covered in this chapter

- Data distribution
- Scaling
- Correlations – simple and multiple
- Redundancy
- Variable selection strategies

3.1 INTRODUCTION

Chapter 1 discussed some important properties of data and Chapter 2 introduced the concept of experimental design and it's application to the selection of different types of data sets. One of these is the training or learning set which is used to produce some mathematical or statistical model or no model at all when used in an unsupervised learning method (see Chapters 4 and 5). Another is the test or prediction set which is used to assess the usefulness of some fitted model and sometimes an evaluation set which has played no part in creating or testing the model. An evaluation set is often a set of data which was deliberately held back from the investigator or which became available after the model or models were constructed. Unfortunately, there is no standard for this nomenclature. An evaluation set, which may also be called a control set, can be used to judge an early stopping point when training an artificial neural network

A Practical Guide to Scientific Data Analysis David Livingstone
© 2009 John Wiley & Sons, Ltd

(see Section 9.3.3). In this case the test set contains samples which have played no part in creating the model.

So far so good, but what about the variables themselves? Do these need to be selected and is there anything that needs to be done with them before an analysis is started? The answer to both these questions is yes and the correct pre-treatment and variable selection may be crucial to the successful outcome of an analysis as the next few sections will show.

3.2   DATA DISTRIBUTION

As discussed in Section 1.4.2 of Chapter 1, knowledge of the distribution of the data values of a variable is important. Examining the distribution allows the identification of outliers, whether real or artefacts, shows whether apparently continuous variables really are and gives an idea of how well the data conforms to the assumptions (usually of normality) employed in some analytical methods. A prime example of this is the very commonly used technique of regression (see Chapter 6) which depends on numerous assumptions about the distribution of the data. Thus, it is often rewarding to plot data values and examine the frequency distributions of variables as a preliminary step in any data analysis. This, of course, is easy to do if the problem consists of only one or two dependent variables and a few tens of independent variables but becomes much more tedious if the set contains hundreds of variables. This, amongst other reasons, is why it is very rarely done! At the very least one should always examine the distribution of any variables that end up in some sort of statistical or mathematical model of a data set.

One particularly important property of variables that can be easily checked, even when considering a large number of variables, is the variance (s^2) or standard deviation (s) which was discussed in Section 1.4.2. The sample variance is the average of the squares of the distance of each data value from the mean:

$$s^2 = \frac{\sum (X - \overline{X})^2}{n - 1} \tag{3.1}$$

Calculation of the variance and standard deviation for a set of variables is a trivial calculation that is a standard feature of all statistics packages. Examination of the standard deviation (which has the same units as the original variable) will show up variables that have small values, which means variables that contain little information about the samples. In the limit, when the standard deviation of a variable is zero, then that variable contains no information at all since all of the values for every sample are the same. It isn't possible to give any 'guide' value for a 'useful' standard deviation since this depends on the units of the original measurements but if we have a set of, say, 50 variables and need to reduce this to 10 or 20 then a reasonable filter would be to discard those variables with the smallest standard deviations. It goes without saying that variables with a standard deviation of zero are useless!

The mean and standard deviation of a data set are dependent on the first and second 'moments' of the set. The term moments here refers to the powers that the data are raised to in the calculation of that particular statistic. There are two higher moment statistics that may be used to characterize the shape of a distribution – skewness, based on the third moment, and kurtosis based on the fourth moment. Skewness is a measure of how symmetrical a distribution is around it's mean and for a completely 'normal' distribution with equal numbers of data values either side of the mean the value of skewness should be zero (see Figure 1.6 for examples of skewed distributions). Distributions with more data values smaller than the mean are said to be positively skewed and will generally have a long right tail so they are also known as 'skewed to the right'. Negatively skewed distributions have the opposite shape, of course. Kurtosis is a measure of how 'peaked' a distribution is as shown in Figure 1.6. Kurtosis, as measured by the moment ratio, has a value of 3 for a normal distribution, although there are other ways of calculating kurtosis which give different values for a normal distribution. It is thus important to check how kurtosis is computed in the particular statistics package being used to analyse data. Distributions which have a high peak (kurtosis > 3) are known as leptokurtic, those with a flatter peak (kurtosis < 3) are called platykurtic and a normal distribution is mesokurtic.

These measures of the spread and shape of the distribution of a variable allow us to decide how close the distribution is to normal but how important is this and what sort of deviation is acceptable? Perhaps unsurprisingly, there is no simple answer to these questions. If the analytical method which will be used on the data depends on assumptions of normality, as in linear regression for example, then the nearer the distributions are to normal the more reliable the results of the analysis will be. If, however, the analytical technique does not rely on assumptions of normality then deviations from normality may well not matter at all. In any case, any 'real' variable is unlikely to have a perfect, normal distribution. The best use that can be made of these measures of normality is as a filter for the removal of variables in order to reduce redundancy in a data set. As will be seen in later sections, variables may be redundant because they contain the same or very similar information to another variable or combination of variables. When it is necessary to remove one of a pair of variables then it makes sense to eliminate the one which has the least normal distribution.

3.3 SCALING

Scaling is a problem familiar to anyone who has ever plotted a graph. In the case of a graph, the axes are scaled so that the information present in each variable may be readily perceived. The same principle applies to the scaling of variables before subjecting them to some form of analysis. The objective of scaling methods is to remove any weighting which is solely due to the units which are used to express a particular variable. An example of this is measurement of the height and weight of people. Expressing height in feet and weight in stones gives comparable values but inches and stones or feet and pounds will result in apparent greater emphasis on height or weight, respectively. Another example can be seen in the values of ^{1}H and ^{13}C NMR shifts. In any comparison of these two types of shifts the variance of the ^{13}C measurements will be far greater simply due to their magnitude. One means by which this can be overcome, to a certain extent at least, is to express all shifts relative to a common structure, the least substituted member of the series, for example. This only partly solves the problem, however, since the magnitude of the Δ shifts will still be greater for ^{13}C than for ^{1}H. A commonly used steric parameter, MR, is often scaled by division by 10 to place it on a similar scale to other parameters such as π and σ.

These somewhat arbitrary scaling methods are far from ideal since, apart from suffering from subjectivity, they require the individual inspection of each variable in detail which can be a time-consuming task. What other forms of scaling are available? One of the most familiar is called *normalization* or *range scaling* where the minimum value of a variable is set to zero and the values of the variable are divided by the range of the variable

$$X'_{ij} = \frac{X_{ij} - X_j\,(\text{MIN})}{X_j\,(\text{MAX}) - X_j\,(\text{MIN})} \qquad (3.2)$$

In this equation X'_{ij} is the new range-scaled value for row i (case i) of variable j. The values of range-scaled variables fall into the range

$0 =< X_j =< 1$; the variables are also described as being normalized in the range zero to one. Normalization can be carried out over any preferred range, perhaps for aesthetic reasons, by multiplication of the range-scaled values by a factor. A particular shortcoming of range scaling is that it is dependent on the minimum and maximum values of the variable, thus it is very sensitive to outliers. One way to reduce this sensitivity to outliers is to scale the data by subtracting the mean from the data values, a process known as mean centring:

$$X'_{ij} = X_{ij} - \overline{X}_j \tag{3.3}$$

As for Equation (3.2), X'_{ij} is the new mean-scaled value for row i (case i) of variable j where $\overline{X}_j$ is the mean of variable j. Mean centred variables are better 'behaved' in terms of extreme values but they are still dependent on their units of measurement.

Another form of scaling which is less sensitive to outliers and which addresses the problem of scaling is known as *autoscaling* in which the mean is subtracted from the variable values and the resultant values are divided by the standard deviation

$$X'_{ij} = \frac{X_{ij} - \overline{X}_j}{\sigma_j} \tag{3.4}$$

Again, in this equation X'_{ij} represents the new autoscaled value for row i of variable j, $\overline{X}_j$ is the mean of variable j, and σ_j is the standard deviation given by Equation (3.6).

$$\sigma_j = \sqrt{\left(\sum_{i=1}^{N} \frac{(x_{ij} - \overline{x}_j)^2}{N-1} \right)} \tag{3.5}$$

Autoscaled variables have a mean of zero and a variance (standard deviation) of one. Because they are mean centred, they are less susceptible to the effects of compounds with extreme values. That they have a variance of one is useful in variance-related methods (see Chapters 4 and 5) since they each contribute one unit of variance to the overall variance of a data set. Autoscaled variables are also known as Z scores, symbol Z, or standard scores.

One further method of scaling which may be employed is known as feature weighting where variables are scaled so as to enhance their effects in the analysis. The objective of feature weighting is quite opposite to that of 'equalization' scaling methods described here; it is discussed in detail in Chapter 7.

3.4   CORRELATIONS

When a data set contains a number of variables which describe the same samples, which is the usual case for most common data sets, then some of these variables will have values which change in a similar way across the set of samples. As was shown in the box in chapter two, the way that two variables are distributed about their means is given by a quantity called covariance:

$$C_{(x,y)} = \sum_{i=1}^{n} (x_i - \overline{x})(y_i - \overline{y})/n \qquad (3.6)$$

Where the covariance is positive the values of one variable increase as the values of the other increase, where it is negative the values of one variable get larger as the values of the other get smaller. This can be handily expressed as the correlation coefficient, r shown in Equation (3.7), which ranges from -1, a perfect negative correlation, through 0, no correlation, to $+1$, a perfect positive correlation.

$$r = C_{(x,y)} / [V_{(x)} \times V_{(y)}]^{\frac{1}{2}} \qquad (3.7)$$

If two variables are perfectly correlated, either negatively or positively, then it is clear that one of them is redundant and can be excluded from the set without losing any information, but what about the situation where the correlation coefficient between a pair of variables is less than one but greater than zero? One useful property of the correlation coefficient is that it's square multiplied by 100 gives the percentage of variance that is shared or common to the two variables. Thus, a correlation coefficient of 0.7 between a pair of variables means that they share almost half (49 %) of their variance. A correlation coefficient of 0.9 means a shared variance of 81 %. A diagrammatic representation of these correlations is shown in Figure 3.1.

   These simple, pairwise, correlation coefficients can be rapidly computed and displayed in a table called the correlation matrix, as shown in the next section. Inspection of this matrix allows the ready identification

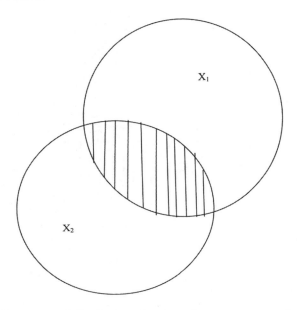

Figure 3.1 Illustration of the sharing of variance between two correlated variables. The hatched area represents shared variance.

of variables that are correlated, a situation also known as *collinearity*, and thus are good candidates for removal from the set.

If two variables can share variance then is it possible that three or more variables can also share variance? The answer to this is yes and this is a situation known as *multicollinearity*. Figure 3.2 illustrates how three variables can share variance.

There is a statistic to describe this situation which is equivalent to the simple correlation coefficient, r, which is called the multiple correlation coefficient. This is discussed further in the next section and in Chapter 6 but suffice to say here that the multiple correlation coefficient can also be used to identify redundancy in a data set and can be used as a criterion for the removal of variables.

3.5 DATA REDUCTION

This chapter is concerned with the pre-treatment of data and so far we have discussed the properties of the distribution of data, means by which data may be scaled and correlations between variables. All of these matters are important, in so far as they dictate what can be done with data,

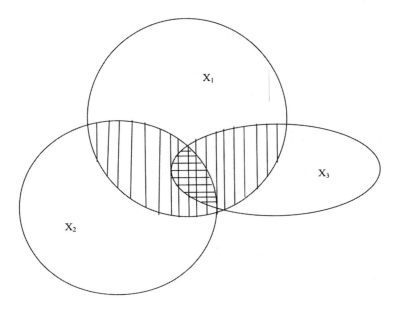

Figure 3.2 Illustration of the sharing of variance between three correlated variables. The hatched areas show shared variance between X_1 and X_2 and X_1 and X_3. The crosshatched area shows variance shared by all three variables.

but perhaps the most important is to answer the question 'What information does the data contain?' It is most unlikely that any given data set will contain as many pieces of information as it does variables.[1] That is to say, most data sets suffer from a degree of redundancy particularly when they contain more variables than cases, a situation in which the data matrix is referred to as being *over-square*. Most people are aware of the fact that with two data points it is possible to construct a line, a 1 dimensional object, and with three data points a plane, a 2 dimensional object. This can be continued so that 4 data points allows a 3 dimensional object, 5 points, 4 dimensions and so on. Thus, the maximum dimensionality of an object, and hence the maximum number of dimensions, in a data set is $n - 1$ where n is the number of data points. For dimensions we can substitute 'independent pieces of information' and thus the maximum that any data set may contain is $n - 1$. This, however, is a *maximum* and in reality the true dimensionality, where dimension means 'information', is often much less than $n - 1$.

---

[1] An example where this is not true is the unusual situation where all of the variables in the set are orthogonal to one another, e.g. principal components (see Chapter 4), but even here some variables may not contain information but be merely 'noise'.

This section describes ways by which redundancy may be identified and, to some extent at least, eliminated. This stage in data analysis is called data reduction in which selected variables are removed from a data set. It should not be confused with dimension reduction, described in the next chapter, in which high-dimensional data sets are reduced to lower dimensions, usually for the purposes of display.

An obvious first test to apply to the variables in a data set is to look for missing values; is there an entry in each column for every row? What can be done if there are missing values? An easy solution, and often the best one, is to discard the variable but the problem with this approach is that the particular variable concerned may contain information that is useful for the description of the dependent property. Another approach which has the advantage of retaining the variable is to delete samples with missing values. The disadvantage of this is that it reduces the size and variety of the data set. In fact either of these methods of dealing with missing data, variable deletion or sample deletion, result in a smaller data set which is likely to contain less information.

An alternative to deletion is to provide the missing values, and if these can be calculated with a reasonable degree of certainty, then all is well. If not, however, other methods may be sought. Missing values may be replaced by random numbers, generated to lie in the range of the variable concerned. This allows the information contained in the variable to be used usefully for the members of the set which have 'real' values, but, of course, any correlation or pattern involving that variable does not apply to the other members of the set. A problem with *random fill* is that some variables may only have certain values and the use of random numbers, even within the range of values of the variable, may distort this structure. In this case a better solution is to randomly take some of the existing values of the variable for other cases and use these to replace the missing values. This has the advantage that the distribution of the variable is unaltered and any special properties that it has, like only being able to take certain values, is unchanged.

An alternative to random fill is *mean fill* which, as the name implies, replaces missing values by the mean of the variable involved. This, like random fill, has the advantage that the variable with missing values can now be used; it also has the further advantage that the distribution of the variable will not be altered, other than to increase its kurtosis, perhaps. Another approach to the problem of missing values is to use linear combinations of the *other* variables to produce an estimate for the missing variable. As will be seen later in this section, data sets sometimes suffer from a condition known as multicollinearity in which one variable is correlated with a linear combination of the other variables. This method of filling missing values certainly involves more work, unless the statistics package has it 'built in', and is probably of debatable value since multicollinearity is a condition which is generally best avoided. There are a number of other ways in which missing data can be filled in and some statistics packages have procedures to analyse 'missingness' and offer a variety of options to estimate the missing values. The ideal solution to missing values, of course, is not to have them in the first place!

The next stage in data reduction is to examine the distribution of the variables as discussed in Section 3.2. A fairly obvious feature to look for in the distribution of the variables is to identify those parameters which have constant, or nearly constant, values. Such a situation may arise because a property has been poorly chosen in the first place, but may also happen when structural changes in the compounds in the set lead to compensating changes in physicochemical properties. Some data analysis packages have a built-in facility for the identification of such ill-conditioned variables. At this stage in data reduction it is also a good idea to actually plot the distribution of each of the variables in the set so as to identify outliers or variables which have become 'indicators', as discussed in Section 1.4.3.

This introduces the correlation matrix. Having removed ill-conditioned variables from the data set, a correlation matrix is constructed by calculation of the correlation coefficient (see Section 3.4) between each pair of variables in the set. A sample correlation matrix is shown in Table 3.1 where the correlation between a pair of variables is found by the intersection of a particular row and column, for example, the correlation between ClogP and Iy is 0.503. The diagonal of the matrix consists of 1.00s, since this represents the correlation of each variable with itself, and it is usual to show only half of the matrix since it is

Table 3.1 Correlation matrix for a set of physicochemical properties.

	Ix	Iy	ClogP	CMR	CHGE(4)	ESDL(4)	DIPMOM	EHOMO
Ix	1.00							
Iy	0.806	1.00						
ClogP	0.524	0.503	1.00					
CMR	0.829	0.942	0.591	1.00				
CHGE(4)	0.344	0.349	0.286	0.243	1.00			
ESDL(4)	0.299	0.257	0.128	0.118	0.947	1.00		
DIPMOM	0.337	0.347	0.280	0.233	0.531	0.650	1.00	
EHOMO	0.229	0.172	0.209	0.029	0.895	0.917	0.433	1.00

symmetrical (the top-right hand side of the matrix is identical to the bottom left-hand side).

Inspection of the correlation matrix allows the identification of pairs of correlating features, although choice of the level at which correlation becomes important is problematic and dependent to some extent on the requirements of the analysis. There are a number of high correlations ($r > 0.9$) in Table 3.1 however, and removal of one variable from each of these pairs will reduce the size of the data set without much likelihood of removing useful information. At this point the data reduction process might begin to be called 'variable selection' which is not just a matter of semantics but actually a different procedure with different aims to data reduction. There are a number of strategies for variable selection; some are applied before any further data analysis, as discussed in the next section, while others are actually an integral part of the data analysis process.

So, to summarize the data reduction process so far:

- Missing values have been identified and the problem treated by either filling them in or removing the offending cases or variables.
- Variables which are constant or nearly constant have been identified and removed.
- Variables which have 'strange' or extreme distributions have been identified and the problem solved, by fixing mistakes or removing samples, or the variables removed.
- Correlated variables have been identified and marked for future removal.

3.6   VARIABLE SELECTION

Having identified pairs of correlated variables, two problems remain in deciding which one of a pair to eliminate. First, is the correlation 'real', in other words, has the high correlation coefficient arisen due to a true correlation between the variables or, is it caused by some 'point and cluster effect' (see Section 6.2) due to an outlier. The best, and perhaps simplest way to test the correlation is to plot the two variables against one another; effects due to outliers will then be apparent. It is also worth considering whether the two parameters are likely to be correlated with one another. In the case of molecular structures, for example, if one descriptor is electronic and the other steric then there is no reason to expect a correlation, although one may exist, of course. On the other hand, maximum width and molecular weight may well be correlated for a set of molecules with similar overall shape.

The second problem, having decided that a correlation is real, concerns the choice of which descriptor to eliminate. One approach to this problem is to delete those features which have the highest number of correlations with other features. This results in a data matrix in which the maximum number of parameters has been retained but in which the inter-parameter correlations are kept low. Another way in which this can be described is to say that the correlation structure of the data set has been simplified. An alternative approach, where the major aim is to reduce the overall size of a data set, is to retain those features which correlate with a large number of others and to remove the correlating descriptors.

Which of these two approaches is adopted depends not only on the data set but also on any knowledge that the investigator has concerning the samples, the dependent variable(s) and the independent variables. In the case of molecular design it may be desirable to retain some particular descriptor or group of descriptors on the basis of mechanistic information or hypothesis. It may also be desirable to retain a descriptor because we have confidence in our ability to predict changes to its value with changes in chemical structure; this is particularly true for some of the more 'esoteric' parameters calculated by computational chemistry techniques. What of the situation where there is a pair of correlated parameters and each is correlated with the same number of other features? Here, the choice can be quite arbitrary but one way in which a decision can be made is to eliminate the descriptor whose distribution deviates most from normal. This is used as the basis for variable choice in a published procedure for parameter deletion called CORCHOP [1]; a flow chart for this routine is shown in Figure 3.3.

Although the methods which will be used to analyse a data set once it has been treated as described here may not depend on distributional assumptions, deviation from normality is a reasonable criterion to apply. Interestingly, some techniques of data analysis such as PLS (see Chapter 7) depend on the correlation structure in a data set and may appear to work better if the data is not pre-treated to remove correlations. For ease of interpretation, and generally for ease of subsequent handling, it is recommended that at least the very high correlations are removed from a data matrix.

Another source of redundancy in a data set, which may be more difficult to identify, is where a variable is correlated with a linear combination of two or more of the other variables in the set. This situation is known

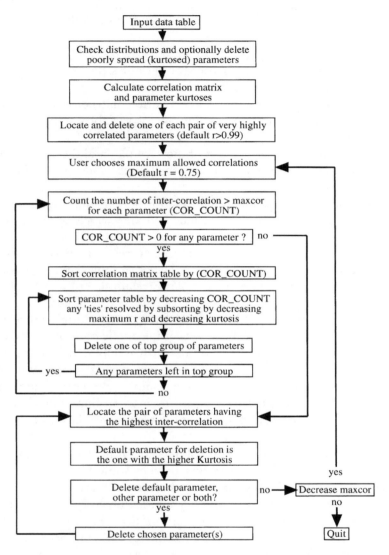

Figure 3.3 Flow diagram for the correlation reduction procedure CORCHOP (reproduced from ref. [2] with permission of Wiley-VCH).

as multicollinearity and may be used as a criterion for removing variables from a data set as part of data pre-treatment. An example of the use of multicollinearity for variable selection is seen in a procedure [2] called UFS (Unsupervised Forward Selection) which is available from the website of the Centre for Molecular Design (www.cmd.port.ac.uk). UFS constructs a dataset by selecting variables with low multicollinearity in the following way:

- The first step is the elimination of variables that have a standard deviation below some assigned lower limit.
- The algorithm then computes a correlation matrix for the remaining set of variables and chooses the pair of variables with the lowest correlation.
- Correlations between the rest of the variables and these two chosen descriptors are examined and any that exceed some pre-set limit are eliminated.
- Multiple correlations between each of the remaining variables and the two selected ones are examined and the variable with the lowest multiple correlation is chosen.
- The next step is to examine multiple correlations between the remaining variables and the three selected variables and to select the descriptor with the lowest multiple correlation.

This process continues until some predetermined multiple correlation coefficient limit is reached. The results of the application of CORCHOP and UFS can be quite different as the former only considers pairwise correlations. The aim of the CORCHOP process is to simplify the correlation structure of the data set while retaining the largest number of descriptors. The aim of the UFS procedure is to produce a much simplified data set in which both pairwise and multiple correlations have been reduced.

It is desirable to remove multicollinearity from data sets since this can have adverse effects on the results given by some analytical methods, such as regression analysis (Chapter 6). Factor analysis (Chapter 5) is one method which can be used to identify multicollinearity. Finally, a note of caution needs to be sounded concerning the removal of descriptors based on their correlation with other parameters. It is important to know which variables were discarded because of correlations with others and, if possible, it is best to retain the original starting data set. This may seem like contrary advice since the whole of this chapter has dealt with the matter of simplifying data sets and removing redundant information. However, consider the situation where two variables have a correlation coefficient of 0.7. This represents a shared variance of just under 50 %, in other words each variable describes just about half of the information in the other, and this might be a good correlation coefficient cut-off limit for removing variables. Now the correlation coefficient between two

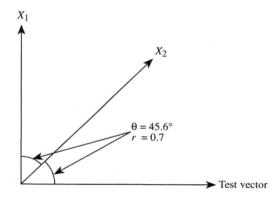

Figure 3.4 Illustration of the geometric relationship between vectors and correlation coefficients (reproduced from ref. [2] with permission of Wiley-VCH).

parameters also represents the angle between them if they are considered as vectors, as shown in Figure 3.4.

A correlation coefficient of 0.7 is equivalent to an angle of approximately 45°. If one of the pair of variables is correlated with a dependent variable with a correlation coefficient of 0.7 this may well be very useful in the description of the property that we are interested in. If the variable that is retained in the data set from that pair is one that correlates with the dependent (X_1 in Figure 3.4) then all is well. If, however, X_1 was discarded and X_2 retained then this parameter may now be completely uncorrelated ($\theta = 90°$) with the dependent variable. Although this is an idealized case and perhaps unlikely to happen so disastrously in a multivariate data set, it is still a situation to be aware of. One way to approach this problem is to keep a list of all the sets of correlated variables that were in the starting set. Figure 3.5 shows a diagram of the correlations between a set of parameters before and after treatment with the CORCHOP procedure. In this figure the correlation between variables is given by the similarity scale. The correlation between LOGPRED and Y_PEAX, for example, is just over 0.4. It can be seen from the figure that there were 4 other variables with a correlation of ~0.8 with LOGPRED (shown by dotted lines in the adjacent cluster) which have been eliminated by the CORCHOP algorithm. If no satisfactory correlations with activity are found in the de-correlated set, individual variables can be re-examined using a diagram such as Figure 3.5. A list of such correlations may also assist when attempts are made to 'explain' correlations in terms of mechanism or chemical features.

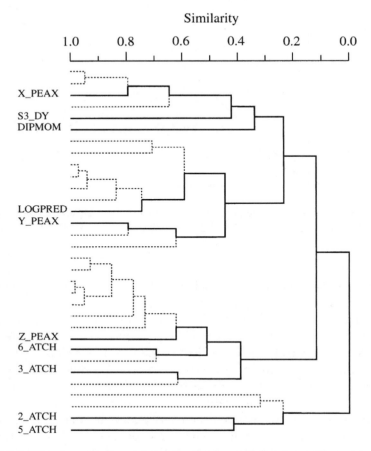

Figure 3.5 Dendrogram showing the physicochemical descriptors (for a set of anti-malarials) retained after use of the CORCHOP procedure. Dotted lines indicate parameters that were present in the starting set (reproduced from ref. [3] with permission of Wiley-Blackwell).

A recent review treats the matter of variable selection in some detail [4].

3.7 SUMMARY

Selection of the analytical tools, described in later chapters, which will be used to investigate a set of data should not be dictated by the availability of software on a favourite computer, by what is the current trend, or by personal preference, but rather by the nature of the data within the set.

The statistical distribution of the data should also be considered, both when selecting analytical methods to use and when attempting to interpret the results of any analysis. The first stage in data analysis, however, is a careful examination of the data set. It is important to be aware of the scales of measurement of the variables and the properties of their distributions (Chapter 1). The cases (samples, objects, compounds, etc.) need selection for the establishment of training and test sets (Chapter 2) although this may have been done at the outset before the data set was collected. Finally, the variables need examination so that ill-conditioned variables can be removed, missing values identified and treated, redundancy reduced and possibly variables selected. All of this is known as pre-treatment and is necessary in order to give the data analysis methods a good chance of success in extracting information.

In this chapter the following points were covered:

1. how an examination of the distribution of variables allows the identification of variables to remove;
2. the need for scaling and examples of scaling methods;
3. ways to treat missing data;
4. the meaning and importance of correlations between variables both simple and multiple;
5. the reasons for redundancy in a data set;
6. the process of data reduction;
7. procedures for variable selection which retain the maximum number of variables in the set or which result in a data set containing minimum multicollinearity.

REFERENCES

[1] Livingstone, D.J. and Rahr, E. (1989). *Quantitative Structure–Activity Relationships*, 8, 103–8.
[2] Whitley, D.C., Ford, M.G. and Livingstone, D.J. (2000). *Journal of Chemical Information and Computer Science*, 40, 1160–1168.
[3] Livingstone, D.J. (1989). *Pesticide Science*, 27, 287–304.
[4] Livingstone, D.J. and Salt, D.W. (2005). Variable Selection – Spoilt for Choice? in *Reviews in Computational Chemistry*, Vol 21, K. Lipkowitz, R. Larter and T.R. Cundari (Eds), pp. 287–348, Wiley-VCH.

4

Data Display

Points covered in this chapter

- Variable by variable plots
- Principal component analysis
- Principal component scores and loadings plots
- Nonlinear mapping
- Artificial neural network plots
- Faces, flower plots, etc

4.1 INTRODUCTION

This chapter is concerned with methods which allow the display of data. The old adage 'a picture is worth a thousand words' is based on our ability to identify visual patterns; it is probably true to say that man is the best pattern-recognition machine that we know of. Unfortunately, we are at our best when operating in only two or three dimensions, although it might be argued that we do operate in higher dimensions if we consider the senses such as taste, smell, touch, and, perhaps, the dimension of time. There are a number of techniques which can help in trying to 'view' multidimensional data and it is perhaps worth pointing out here that this is exactly what the methods do – they allow us to view a data set from a variety of perspectives. If we consider a region of attractive countryside, or a piece of famous architecture such as the Taj Mahal, there is no 'correct' view to take of the scene. There are, however, some views which are 'better' from the point of view of an appreciation

A Practical Guide to Scientific Data Analysis   David Livingstone
© 2009 John Wiley & Sons, Ltd

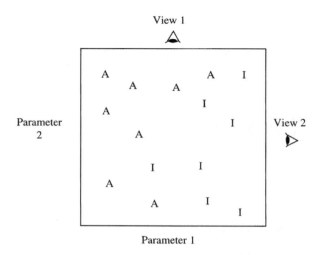

Figure 4.1 Plot of a set of active (A) and inactive (I) compounds described by two physicochemical properties.

of the beauty of the scene, a view of the Taj Mahal which includes the fountains, for example. Figure 4.1 shows a plot of the values of two parameters against one another for a set of compounds which are marked as A for active and I for inactive.

Looking at the plot as presented gives a clear separation of the two classes of compounds, the view given by the two parameters is useful. If we consider the data represented by parameter 1, seen from the position marked view 1, it is seen that this also gives a reasonable separation of the two classes although there is some overlap. The data represented by parameter 2 (view 2), on the other hand, gives no separation of the classes at all. This illustrates two important features. First, the consideration of multiple variables often gives a better description of a problem: in this case parameter 2 helps to resolve the conflict in classification given by parameter 1. Second, the choice of viewpoint can be critical and it is usually not possible to say in advance what the 'best' viewpoint will be. Hopefully this simple two-dimensional example has illustrated the problems that may be encountered when viewing multivariate data of 50, 100, or even more dimensions.

Now to multivariate display methods. These methods can conveniently be divided into linear and nonlinear techniques as discussed in the next two sections; cluster analysis as a display method is covered in Chapter 5.

4.2   LINEAR METHODS

The simplest and most obvious linear display method is a variable-by-variable plot. The advantages of such plots are that they are very easy to interpret and it is easy to add a new point to the diagram for prediction or comparison (this is not necessarily the case with other methods, as will be shown later). One of the disadvantages of such an approach is that for a multivariate set, there can be many two-dimensional plots, $p(p-1)$ for p variables. Such plots not only take time to generate but also take a lot of time to evaluate. Another disadvantage of this technique is the limited information content of the plots; Figure 4.1 shows the improvement that can be obtained by the addition of just one parameter to a variable which already describes the biological data reasonably well. How can further dimensions be added to a plot? Computer graphics systems allow the production of three-dimensional pictures which can be viewed in stereo and manipulated in real time. They are often used to display the results of molecular modelling calculations for small molecules and proteins, but can just as easily be adapted to display data. The advantage of being able to manipulate such a display is that a view can be selected which gives the required distribution of data points; in Figure 4.1, for example, the best view is above the plot. The use of colour or different-shaped symbols can also be used to add extra dimensions to a plot. Figure 4.2 shows a physical model, a reminder of more 'low-tech' times, in which a third parameter is represented by the height of the columns above the baseboard and activity is represented by colour.

Another approach is shown in the spectral diagram in Figure 4.3 which represents a simultaneous display of the activities of compounds (circles) and the relationships between tests (squares); the areas of the symbols represent the mean activity of the compounds and tests. A fuller description of spectral map analysis is given in Chapter 8.

Through these ingenious approaches it is possible to expand diagrams to four, five, or even six dimensions, but this does not even begin to solve the problem of viewing a 50-dimensional data set. What is required is some method to reduce the dimensionality of the data set while retaining its information content. One such technique is known as *principal component analysis* (PCA) and since it forms the basis of a number of useful methods, both supervised and unsupervised, I will attempt to explain it here in some detail. The following description is based, with very grateful permission, on part of Chapter 6 of the book by Hilary Seal [2].

Figure 4.2 A physical model used to represent three physicochemical properties, π and σ on the baseboard and MR as the height of the balls. Five colours were used to code the balls (representing compounds) for five activity classes.

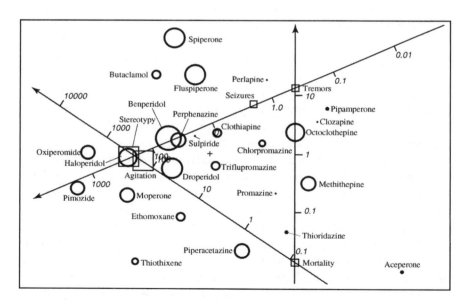

Figure 4.3 Spectral map of the relationships between the activity of neuroleptics (circles) and *in vivo* tests (squares) (reproduced from ref. [1] copyright (1986) Elsevier).

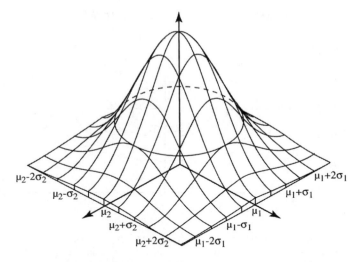

Figure 4.4 Three-dimensional plot of the frequency distribution for two variables. The two variables have the same standard deviations so the frequency 'surface' is symmetrical (reproduced from ref. [2] copyright Methuen).

We have seen in Chapter 1 (Figures 1.3 and 1.4) that a frequency distribution can be constructed for a single variable in which the frequency of occurrence of variable values is plotted against the values themselves. If we take the values of two variables which describe a set of samples (compounds, objects, mixtures, etc.) a frequency distribution can be shown for both variables simultaneously (Figure 4.4).

In this diagram the height of the surface represents the number of occurrences of samples which have variable values corresponding to the X and Y values of the plane which the surface sits on. The highest point of this surface, the summit of the hill, corresponds to the mean of each of the two variables. It is possible to take slices through a solid object such as this and plot these as ellipses on a two-dimensional plot as shown in Figure 4.5. These ellipses represent population contours: as the slices are taken further down the surface from the summit, they produce larger ellipses which contain higher proportions of the population of variable values.

Two important things can be seen from this figure. First, the largest axis of the ellipses corresponds to the variable (X_1) with the larger standard deviation. Thus, the greatest part of the shape of each ellipse is associated with the variable which contains the most variance, in other words, information. Second, the two axes of the ellipses are aligned with the two axes of the plot. This is because the two variables are not

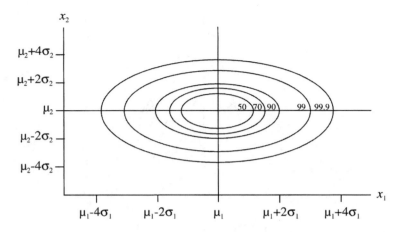

Figure 4.5 Population contours from a frequency distribution such as that shown in Figure 4.4. In this case, the variables have different standard deviations ($\sigma_1 > \sigma_2$) so the contours are ellipses (reproduced from ref. [2] copyright Methuen).

associated with one another; where there are high values of variable X_2, there is a spread of values of variable X_1 and vice versa. If the two variables are correlated then the ellipses are tilted as shown in Figure 4.6 where one population contour is plotted for two variables, Y and X, which are positively correlated.

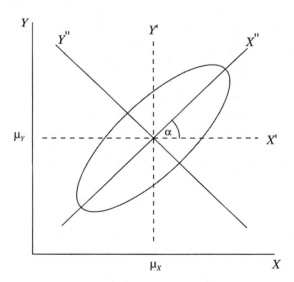

Figure 4.6 Population contour for two correlated variables X and Y. The axes X' and Y' represent mean-centred variables achieved by translation of the origin of X and Y. The axes X'' and Y'' are formed by rotation of X' and Y' through the angle α (reproduced from ref. [2] copyright Methuen).

The two 'ends' of the population ellipse are located in two quadrants of the $X–Y$ space which correspond to (low X, low Y) and (high X, high Y). If the variables were negatively correlated, the ellipse would be tilted so that high values of Y correspond to low values of X and the other end of the ellipse would be in the (low Y, high X) quadrant. The relationship between the two axes, X and Y, and the population ellipse, which can be thought of as enclosing the cloud of data points, shows how well the original variables describe the information in the data. Another way of describing the data is to transform the axes X and Y to new axes X' and Y' as shown in the figure. This is achieved by translation of the origin of X and Y to a new position in the centre of the ellipse, a procedure called, unsurprisingly, centring. A further operation can be carried out on the new axes, X' and Y', and that is rotation through the angle α marked on the figure, to give yet another set of axes, X'' and Y''. These are the two basic operations involved in the production of *principal components*, translation and rotation.

Now it may seem that this procedure has not achieved very much other than to slightly alter two original variables, and both by the same amount, but it will be seen to have considerable effects when we involve larger numbers of variables. For the present, though, consider the results of this procedure as it illustrates some important features of PCA. The new variable, X'', is aligned with the major axis of the ellipse and it is thus explaining the major part of the variance in the data set. The other new variable, Y'', is aligned with the next largest axis of the ellipse and is thus explaining the next largest piece of information in the set of data points. Why is this the next largest piece of variance in the data set; surely another direction can be found in the ellipse which is different to the major axis? The answer to this question is yes, but a requirement of principal components is that they are orthogonal to one another (also uncorrelated) and in this two-dimensional example that means at 90°. The two important properties of principal components are:

1. the first principal component explains the maximum variance in the data set, with subsequent components describing the maximum part of the remaining variance subject to the condition that
2. all principal components are orthogonal to one another.

In this simple two-dimensional example it is easy to see the directions that the two principal components (PCs) must take to describe the variance in the data set. Since the two axes, X and Y, were originally orthogonal it is also easy to see that it is only necessary to apply the same single rotation to each axis to produce the PCs. In the situation where there are multiple variables, the same single rotation (including reflection) is applied to all the variables. The other feature of principal components analysis that this example demonstrates is the matter of dimensionality. The maximum number of components which are orthogonal and that can be generated in two dimensions is two. For three dimensions, the maximum number of orthogonal components is three, and so on for higher dimensional data sets. The other 'natural' limit for the number of components that can be extracted from a multidimensional data set is dictated by the number of data points in the set. Each PC must explain some part of the variance in the data set and thus at least one sample point must be associated with each new PC dimension. The third condition for PCA is thus

3. as many PCs may be extracted as the smaller of n (data points) or p (dimensions) for a $n \times p$ matrix (denoted by q in Equation (4.1)).[1]

There are other important properties of PCs to consider, such as their physical meaning and their 'significance'. These are discussed further in this section and in Chapter 7; for the present it is sufficient to regard them as means by which the dimensionality of a high-dimensional data space can be reduced. How are they used? In the situation where a data set contains many variables the PCs can be regarded as new variables created by taking a linear combination of the original variables as shown in Equation (4.1).

$$PC_1 = a_{1,1}v_1 + a_{1,2}v_2 + \ldots a_{1,p}v_p$$
$$PC_2 = a_{2,1}v_1 + a_{2,2}v_2 + \ldots a_{2,p}v_p$$
$$\ldots\ldots\ldots\ldots\ldots\ldots\ldots\ldots\ldots\ldots\ldots\ldots\ldots$$
$$PC_q = a_{q,1}v_1 + a_{q,2}v_2 + \ldots a_{q,p}v_p \qquad (4.1)$$

Where the subscripted term, a_{ij}, represents the contribution of each original variable ($v_1 \rightarrow v_p$) in the P-dimensional set to the particular principal component ($1 \rightarrow q$) where q (the number of principal components) is the smaller of the n points or p dimensions. These coefficients have a sign associated with them, indicating whether a particular variable makes a negative or positive contribution to the component, and their magnitude

---

[1] Actually, it is the rank of the matrix, denoted by $r(A)$, which is the maximum number of linearly independent rows (or columns) in A. $0 \leq r(A) \leq \min (n,p)$, where A has n rows and p columns.

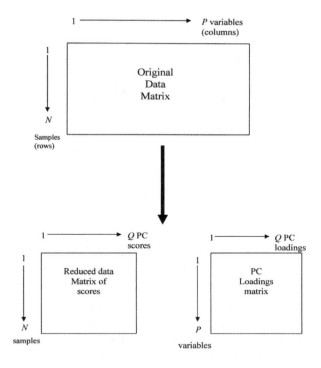

Figure 4.7 Illustration of the process of principal components analysis to produce a 'new' data matrix of Q scores for N samples where Q is equal to (or less than) the smaller of P (variables) or N (samples). The loadings matrix contains the contribution (loading) of each of the P variables to each of the Q principal components.

shows the degree to which they contribute to the component. The coefficients are also referred to as loadings and represent the contribution of individual variables to the principal components.[2] Since the PCs are new variables it is possible to calculate values for each of these components for each of the objects (data points) in the data set to produce a new (reconstructed but related) data set. The numbers in this data set are known as principal component scores; the process is shown diagrammatically in Figure 4.7.

Now, it may not seem that this has achieved much in the way of dimension reduction: while it is true that the scores matrix has a 'width' of q this will only be a reduction if there were fewer compounds than variables in the starting data set. The utility of PCA for dimension

---

[2] A loading is actually the product of the coefficient and the eigenvalue of the principal component (a measure of its importance) as described later.

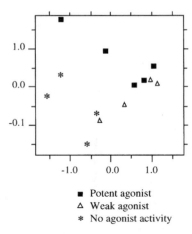

Figure 4.8   Scores plot for 13 analogues of γ-aminobutyric acid (reproduced from ref. [3] with kind permission of Springer Science + Business Media).

reduction lies in the fact that the PCs are generated so that they explain maximal amounts of variance. The majority of the information in many data sets will be contained in the first few PCs derived from the set. In fact, by definition, the most informative view of a data set, in terms of variance at least, will be given by consideration of the first two PCs. Since the scores matrix contains a value for each compound corresponding to each PC it is possible to plot these values against one another to produce a low-dimensional picture of a high-dimensional data set. Figure 4.8 shows a scores plot for 13 compounds described by 33 calculated physicochemical properties.

This picture is drawn from the scores for the first two PCs and it is interesting to see that the compounds are roughly separated into three classes of biological activity – potent, weak, and no agonist activity. Although the separation between classes is not ideal this is still quite an impressive picture since it is an example of unsupervised learning pattern recognition; the biological information was not used in the generation of the PCs. Table 4.1 gives a list of the loadings of the original 33 variables with the first three PCs. This table should give some idea of the complex nature of PCs derived from large dimensional data sets.

Some variables contribute in a negative fashion to the first two PCs, e.g. CMR, 4-ESDL, 3-NSDL, and so on, while others have opposite signs for their loadings on these two PCs. The change in sign for the loadings of an individual variable on two PCs perhaps seems reasonable when

Table 4.1 Loadings of input variables for the first three principal components (total explained variance = 70 %) (reproduced from ref. [3] with kind permission of Springer Science + Business Media).

Variable	Loading		
	PC_1	PC_2	PC_3
CMR	−0.154	−0.275	0.107
1-ATCH	−0.196	0.096	0.298
2-ATCH	0.015	−0.203	−0.348
3-ATCH	0.186	0.003	0.215
4-ATCH	−0.183	0.081	−0.197
5-ATCH	−0.223	0.061	−0.027
6-ATCH	0.272	−0.030	0.049
X-DIPV	0.152	−0.085	−0.059
Y-DIPV	0.079	−0.077	−0.278
Z-DIPV	0.073	−0.117	0.019
DIPMOM	−0.019	0.173	−0.006
T-ENER	0.137	0.146	−0.242
1-ESDL	0.253	−0.156	0.120
2-ESDL	0.221	−0.071	−0.020
3-ESDL	−0.217	0.108	−0.248
4-ESDL	−0.167	−0.115	−0.245
5-ESDL	−0.105	0.197	0.158
6-ESDL	0.128	0.072	0.337
1-NSDL	0.183	−0.253	0.148
2-NSDL	0.186	−0.236	0.025
3-NSDL	−0.021	−0.136	−0.365
4-NSDL	−0.226	0.195	−0.046
5-NSDL	−0.111	0.227	0.141
6-NSDL	−0.099	0.257	0.125
VDW_VOL	−0.228	−0.229	0.031
X-MOFI	−0.186	−0.238	0.136
Y-MOFI	−0.186	−0.266	0.093
Z-MOFI	−0.209	−0.238	0.090
X-PEAX	−0.218	−0.178	−0.020
Y-PEAX	−0.266	−0.050	0.084
Z-PEAX	−0.051	−0.217	0.035
MOL_WT	−0.126	−0.263	0.189
IHET_1	0.185	−0.071	−0.052

we consider that the PCs are orthogonal; the PCs are taking different 'directions' and thus a variable that contributes positively to one PC might be negatively associated with another (see Figure 4.13). Where the signs of the loadings of one variable on two PCs are the same, the loading for that variable on a third PC is often (but not always) reversed, demonstrating that the third component is taking a different direction to the first two. It should be pointed out here that the direction that a PC

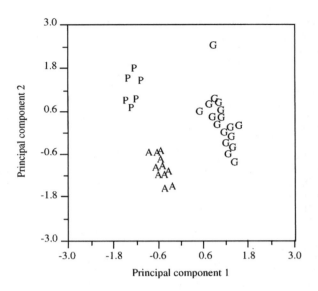

Figure 4.9 Scores plot for fruit juice samples: A, apple juice; P, pineapple juice; and G, grape juice (reproduced from ref. [4] with permission of Wiley-Blackwell).

takes, with respect to the original variables, is arbitrary. Reversing all of the signs of the loadings of the variables on a particular PC produces a component which explains the same amount of variance. When PCs are calculated for the same data set using two different software packages, it is not unusual to find that the signs of the loadings of the variables on corresponding PCs (e.g. the first PC from the two programs) are reversed, but the eigenvalues (variance explained) are the same. The other important piece of information to note in Table 4.1 is the magnitude of the coefficients. Many of the variables that make a large contribution to the first component will tend to have a small coefficient in the second component and vice versa. Some variables, of course, can make a large contribution to both of these PCs, e.g. CMR, T-ENER, 1-ESDL, 1-NSDL, etc., in which case they are likely to make a smaller contribution to the third component. The variable T-ENER demonstrates an exception to this in that it has relatively high loadings on all three components listed in the table.[3]

Figure 4.9 shows an example of the value of PCA in food science. This is a scores plot for the first two PCs derived from a data set of 15

---

[3] For this data set it is possible to calculate a total of 13 components although not all are 'significant' as discussed later.

variables measured on 34 samples of fruit juices. The variables included pH, total phenols, reducing sugars, total nitrogen, ash, glucose content, and formol number, and the samples comprised 17 grape, 11 apple, and six pineapple juices.

As can be seen from the figure, the first two components give a very satisfactory separation of the three types of juice. The first PC was related to richness in sugar since the variables reducing sugars, total sugars, glucose, °Brix, dry extract, and fructose load highly onto it. This component distinguishes grape from apple and pineapple juice. The second PC, which separates apple from pineapple juice was highly correlated with the glucose:fructose ratio, total nitrogen, and formol number. In this example it is possible to attempt to ascribe some chemical 'meaning' to a PC, here, sugar richness described by PC_1, but in general it should be borne in mind that PCs are mathematical constructs without necessarily having any physical significance. An example of the use of PCA in another area of chemical research is shown in Figure 4.10. This scores plot was derived from PCA applied to a set of seven parameters, calculated $\log P$ and six theoretical descriptors, used to describe a series of 14 substituted benzoic acids.

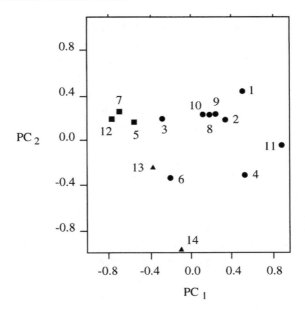

Figure 4.10   Scores plot for a set of benzoic acids described by seven physicochemical properties. Compounds are metabolized by the formation of glucuronide conjugates (squares) or glycine conjugates (circles) and two test set compounds are shown (triangles) (reproduced from ref. [5] with permission of Elsevier).

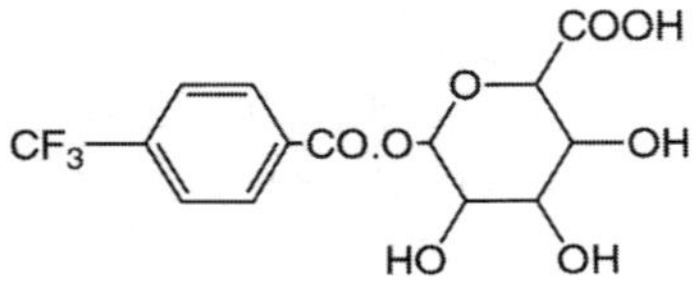

Figure 4.11 Structure of (a) the glycine conjugate of 4-fluorobenzoic acid; (b) the glucuronic acid conjugate of 4-trifluoromethylbenzoic acid. (reproduced from ref. [5] with permission of Elsevier).

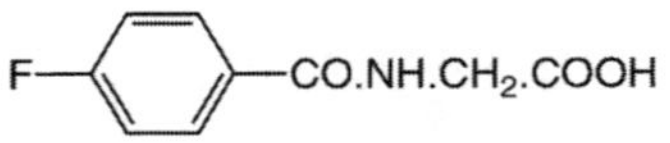

The major route of metabolism of these compounds in the rat was determined by NMR measurements of urine, or taken from the literature, and they were assigned to glycine (Figure 4.11a) or glucuronide (Figure 4.11b) conjugates.

A training set of 12 compounds is shown on the scores plot in Figure 4.10 where it can be seen that the glucuronide conjugate-forming compounds (squares) are well separated from the rest of the set. Two test set compounds are shown as triangles; compound 13 is metabolized by the glucuronide route and does lie close to the other glucuronide conjugate formers. However, this compound is also close, in fact closer, to a glycine conjugate-forming acid (number 6) and thus might be predicted to be metabolized by this route. The other test set compound lies in a region of PC space, low values of PC$_2$, which is not mapped by the other compounds in the training set. This compound is metabolized by the glycine conjugate route but it is clear that this could not be predicted from this scores plot. This example serves to illustrate two points. First, the PC scores plot can be used to classify successfully the metabolic route for the majority of these simple benzoic acid analogues, but that individual predictions may not be unambiguous. Second, it demonstrates the importance of careful choice of test and training set compounds. Compound 14 must have some extreme values in the original data matrix and thus might be better treated as a member of the training set. In fairness to the original report it should be pointed out that the selection of 'better' variables, in terms of their ability to classify the compounds, led to plots with much better predictive ability.

As was seen in Table 4.1, PCA not only provides information about the relationships between samples in a data set but also gives us insight into the relationships between variables. The schematic representation

Table 4.2 Loadings for selected variables on the first two PCs* (reproduced from ref. [6] with kind permission of Springer Science + Business Media).

No. on Figure 4.12	Parameter	PC$_1$	PC$_2$
1	PIAR	−0.72	0.11
2	PIAL	−0.66	0.16
3	FARR	−0.71	0.23
4	FALR	−0.69	0.21
5	FARHL	−0.72	0.17
6	FALHL	0.69	0.23
7	K	0.04	−0.49
38	SX	0.14	0.04
43	RE	0.07	0.01
44	I	0.14	0.07
50	HD	0.12	0.07
59	RAND	0.09	0.39

*From a total of 75 parameters describing 59 substituents.

of PCA in Figure 4.7 shows that the process produces two new matrices, each of width Q, where Q is the smaller of P (variables) or N (samples). The scores matrix contains the values of new variables (scores) which describe the samples. The loadings matrix contains the values of the loadings (correlations) of each variable with each of the Q principal components. These loadings are the coefficients for the variables in Equation (4.1) and can be used to construct a loadings plot for a pair of PCs. In an analysis of an extensive set of physicochemical substituent constants, Van de Waterbeemd and colleagues [6] produced the PC loadings shown in Table 4.2.

The loadings for the full set of substituent constants are shown projected onto the first two PC axes in Figure 4.12. In this figure each point represents a variable; where points are clustered together, the variables are all highly associated with one another. Those points which lie close to the origin of the plot (e.g. 38, 43, 44, and 50) make little contribution to either PC, conversely the points at the extremes of the four quadrants are highly associated with their respective PCs.

The cluster of variables represented by points 1 to 6 is a chemically 'sensible' one, these are all descriptors of lipophilicity. The fact that parameter 59 lies close to the origin is reassuring, this variable was generated from random numbers. Descriptor 7 is derived from measurements of charge-transfer complexes, its relationship to other parameters is examined further in Section 7.3. Points which lie on their own in the PC space represent variables which contain some unique information not associated with other variables.

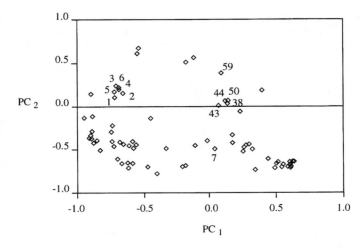

Figure 4.12 Loadings plot for a set of 75 substituent constants (reproduced from ref. [6] with kind permission of Springer Science + Business Media).

By joining the points representing variables to the origin of the PC plot it is possible to construct vectors in the two-dimensional plane of PC space. This type of representation can be adapted to produce a diagram which aims to give another, more visual, explanation of principal component analysis. In Figure 4.13 the solid arrows represent individual

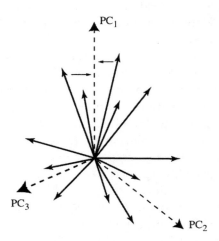

Figure 4.13 Pictorial representation of the relationship between data vectors (variables), shown by solid lines, and PCs shown by dotted lines. The plane of the diagram is not 'real' two-dimensional space or PC space but is meant to represent P dimensions.

variables as vectors, with the length of each arrow proportional to the variance contained in the variable. This is not the same type of plot as Figure 4.12; the two-dimensional space is not PC space but is intended to represent P dimensions.

The position of the arrows in the diagram demonstrates the relationships between the variables, arrows which lie close to one another represent correlated variables. The first PC is shown as a dotted arrow and it can be seen to lie within a cluster of correlated variables. The loadings of these variables (and the others in the set) are found by projection of the arrows onto this PC arrow, illustrated for just two variables for clarity. The length of the PC arrow is given by vector addition of the arrows representing the variables and, as for the individual variables, this represents the variance contained in this component. The second and third PCs lie within other sets of correlated variables and are shorter vectors than the first since they are explaining smaller amounts of variance in the set. The PC vectors are not at right angles (orthogonal to one another) in this diagram since the space in the figure is not 'real' two-dimensional space. The relationship between PC vectors and the variable vectors illustrates an operation that can be carried out on PCs in order to simplify their structure. This can be of assistance in attempts to interpret PCs and may also result in PCs which are better able to explain some dependent variable. The three PC vectors shown in Figure 4.13 were generated so as to explain the maximum variance in the data set and thus there are a lot of variables associated with them. This association of many variables with each component leads to low loadings for some of the variables, particularly some of the more 'important' (high-variance) variables. By trying to explain the maximum amount of variance in the set, PCA achieves a compromise between PC 'directions' that are aligned with high-variance variables and directions that are aligned with a large number of variables. Rotation of the PCs allows new directions to be found in which fewer variables are more highly associated with each PC. There are a number of techniques available to achieve such rotations, one of the commonest is known as *varimax* rotation [7]. Table 4.3 shows the loadings of seven physicochemical parameters on four PCs for a set of 18 naphthalene derivatives. High loadings, i.e., variables making a large contribution, for each component are shown in bold type.

It can be seen that the first component in particular has a quite complicated structure with four variables contributing to it and that two of these, π and MR, are properties that it is desirable to keep uncorrelated. Table 4.4 shows the loadings of these same variables on four PCs after varimax rotation.

Table 4.3 Parameter loadings for four principal components (reproduced from ref. [8] with permission of Elsevier).

Parameter	1	2	3	4
π	0.698*	−0.537	−0.121	−0.258
MR	0.771	0.490	−0.302	−0.002
F	0.261	0.389	0.745	−0.423
R	0.405	−0.012	0.578	0.697
H_a	−0.140	0.951	0.071	−0.101
H_d	−0.733	0.373	−0.271	0.172
$^1\chi_{sub}^v$	0.739	0.412	−0.404	0.163

*Boldface numbers indicate parameters making a large contribution to each component.

The structure of the first PC has been simplified considerably and the correlation between π and MR has been almost eliminated by reducing the π loading from 0.6988 to 0.2. This parameter now loads onto the second PC (note the change in sign) and the properties which were highly associated with the third and fourth PCs have had their loadings increased. Varimax rotation results in a new set of components, often referred to as factors, in which loadings are increased or reduced to give a simplified correlation structure. This rotation is orthogonal, that is to say the resulting factors are orthogonal like the PCs they were derived from. Other orthogonal rotations may be used to aid in the interpretation of PCs and non-orthogonal (oblique) rotations also exist [7].

Scores or loadings plots are not restricted to the first two PCs, although all of the examples shown so far have been based on the first two PCs. By definition, the first two PCs explain the largest amount of variance in a data set, but plots of other components may be more informative; section 7.3.1, for example, shows a data set where the first and fourth PCs were most useful in the explanation of a dependent variable. Plots

Table 4.4 Parameter loadings after varimax rotation (reproduced from ref. [8] with permission of Elsevier).

Parameter	1	2	3	4
π	0.200	0.919*	0.012	0.012
MR	0.891	0.195	0.093	0.061
F	0.020	−0.003	0.975	0.123
R	0.081	0.018	0.115	0.982
H_a	0.272	0.451	0.318	−0.083
H_d	−0.159	−0.285	−0.138	−0.148
$^1\chi_{sub}^v$	0.974	0.037	−0.024	0.064

*Boldface numbers indicate parameters making a large contribution to each component.

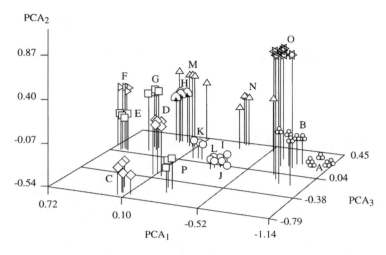

Figure 4.14 Scores plot on the first three PCs for a set of natural orange aroma samples described by GC–MS data. Different samples are indicated by the letters A–P, and different categories by different symbols (reproduced from ref. [9] with permission of Elsevier).

are also not restricted to just two PCs, although two-dimensional plots are quite popular since they fit easily onto two-dimensional paper! The physical model shown earlier (Figure 4.2) is a four-dimensional plot and the spectral map (Figure 4.3) contains a third dimension in the thickness of the symbols. Figure 4.14 shows a plot of the first three PCs calculated from a GC–MS analysis (32 peaks) of natural orange aroma samples. The different samples, labelled A to P, were of distinct types of orange aroma provided by six different commercial flavour houses. These orange aromas could be classified into nine separate categories, as indicated by the different symbols on the plot, and it can be seen that this three-dimensional diagram separates the categories quite well.

The plots shown in these examples have involved either the loadings or the scores from PCA but it is in fact possible to produce plots, known as biplots, which simultaneously display the scores and loadings [10]. The advantage of a biplot is that it enables the analyst to examine relationships between variables and the cases at the same time. The disadvantage, of course, is that it can be more difficult to see any resulting patterns. Biplots are discussed further in Section 8.4.

As mentioned at the beginning of this section, PCA lies at the heart of several analytical methods which will be discussed in later chapters. These techniques such as factor analysis, as shown in the next chapter, and Partial Least Squares (Chapter 7) can be used to produce scores and loadings plots in a similar way to PCA. Some other features of PCs, such as their 'significance', are also discussed later in the book; this section has been intended to illustrate the use of PCA as a linear dimension reduction method.

4.3   NONLINEAR METHODS

The next two sections discuss nonlinear approaches to data display. The first section describes a method in which cases in the data set are displayed in two dimensions while striving to preserve the inter-sample distances as measured in the multidimensional space. There are two techniques for achieving this, based on similar concepts, which are called multidimensional scaling [11, 12] and nonlinear mapping [13, 14]. The section describes the procedure for nonlinear mapping. The second section introduces a data display technique based on an approach called artificial neural networks. The underlying philosophy of artificial neural networks is described in some detail later in the book so this section only briefly discusses the operation of a particular network architecture and algorithm.

4.3.1   Nonlinear Mapping

For any given data set of points in P dimensions it is possible to calculate the distances between pairs of points by means of an equation such as that shown in Equation (4.2).

$$d_{ij} = \sqrt{\left(\sum_{k=1,P} (d_{i,k} - d_{j,k})^2 \right)} \qquad (4.2)$$

This is the expression for the Euclidean distance where d_{ij} refers to the distance between points i and j in a P-dimensional space given by the summation of the differences of their coordinates in each dimension ($k = 1, P$). Different measures of distance may be used to characterize the similarities between points in space, e.g. city-block distances, Mahalonobis distance (see Digby and Kempton [15] for examples), but for most purposes the familiar Euclidean distance is sufficient. The collection of interpoint distances is known, unsurprisingly, as a distance matrix (see Section 5.2 for an example) and this is used as the starting point for a number of multivariate techniques.

The display method considered here is known as *nonlinear mapping*, NLM for short, and takes as its starting point the distance matrix for a data set calculated according to Equation (4.2). The distances in this distance matrix are labelled d_{ij}^* to indicate that they relate to *P*-space interpoint distances. Having calculated the *P*-space distance matrix, the next step is to randomly (usually, but see later) assign the points (compounds, samples) to positions in a lower dimensional space. This is usually a two-dimensional space for ease of plotting but can be a three-dimensional space if a computer is used to rotate the plot to show the third dimension. It could also be a true 3-D display if a computer graphics display with stereo is used, or a two-dimensional stereo plot with appropriate viewer. Having assigned the *n* points to positions in a two-dimensional coordinate system, distances between the points can be calculated using Equation (4.2) and these are labelled d_{ij}. The difference between the *P*-space interpoint distances and the 2-space interpoint distances can be expressed as an error, *E*, as shown in Equation (4.3).

$$E = \sum_{i>j} (d_{ij}^* - d_{ij})^2 / (d_{ij}^*)^\rho \qquad (4.3)$$

Minimization of this error function results in a two-dimensional display of the data set in which the distances between points are such that they best represent the distances between points in *P*-space. The significance of the power term, ρ, will be discussed later in this section; it serves to alter the emphasis on the relative importance of large versus small *P*-space interpoint distances.

A physical analogy of the process of NLM can be given by consideration of a three-dimensional object composed of a set of balls joined together by springs. If the object is pushed onto a flat surface and the tension in the springs allowed to equalize, the result is a two-dimensional representation of a three-dimensional object. The equalization of tension in the springs is equivalent to minimization of the error function in Equation (4.3). A two-dimensional plot produced by the NLM process has some interesting features. Each axis of the plot consists of some (unknown) nonlinear combination of the properties which were used to define the original *P*-dimensional data space. Thus, it is not possible to plot another point directly onto an NLM; the whole map must

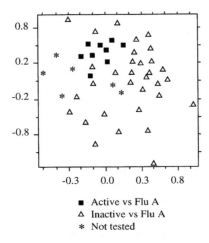

Figure 4.15   NLM of a set of antiviral bicyclic amine derivatives (reproduced from ref. [3] with kind permission of Springer Science + Business Media).

be recalculated with the new point included. An example of an NLM which includes both training set and test set compounds is shown in Figure 4.15. This plot was derived from a set of bicyclic amine derivatives which were described by nine calculated parameters. Antivirus activity results were obtained from a plaque-reduction assay against influenza A virus. It can be seen from the map that the active compounds are grouped together in one region of space. Some of the test set compounds lie closer to this region of the plot, though none of them within it, and thus the expectation is that these compounds are more likely to be active than the other members of the test set.

This is a good example of the use of a NLM as a means for deciding the order in which compounds should be made or tested. Another use for NLM is to show how well physicochemical property space is spanned by the compounds in the training set, or test set for that matter. Regions of space on the NLM which do not contain points probably indicate regions of P-space which do not contain samples. The qualifier 'probably' was used in the last statement because the space on an NLM does not correspond directly to space in P dimensions. A map is produced to meet the criterion of the preservation of interpoint distances so, as we move about in the 2-space of an NLM this might be equivalent to quite strange moves in P-space. Small distances on the NLM may be equivalent to large distances in the space of some variables, small or zero distances with respect to other variables and may even involve a change of direction in the space of some variables.

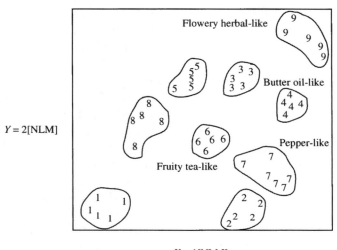

$$X = 1[\text{NLM}]$$

Figure 4.16 NLM of natural orange aroma samples described by 32 GC–MS peaks (reproduced from ref. [9] with permission of Elsevier).

Another example of the use of NLM to treat chemical data is shown in Figure 4.16.

This NLM was calculated from the same GC–MS data used to produce the principal component scores plot shown in Figure 4.14. The NLM clearly groups the samples into nine different categories, the descriptions of the samples are comments made by a human testing panel (see later). Figure 4.17 shows another example of an NLM, this time

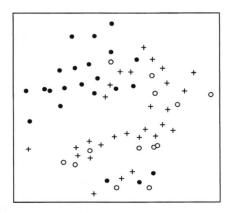

Figure 4.17 NLM of hallucinogenic phenylalkylamine derivatives described by 24 physicochemical properties; • is active, + is low activity, 0 is inactive (from ref. [16] copyright (1990) American Chemical Society).

from the field of drug design. This plot shows 63 hallucinogenic phenylalkylamine derivatives characterized by 24 physicochemical properties.
Compounds with high activity are mostly found in the top-left quadrant
of the map, the inactive and low-activity compounds being mixed in the
rest of the space of the map. Interestingly, this map also shows three
active compounds which are separated from the main cluster of actives.

These compounds lie quite close to the edge of the plot and thus in a
region of the NLM space that might be expected to behave in a pecu-
liar fashion. They may actually be quite similar to the rest of the active
cluster, in other words the map may 'join up' at the axes and they are
simply placed there as a good compromise in the minimization of the
error function. An alternative explanation is that these compounds exert
their activity due to some unique features, they may act by a different
mechanism or perhaps occupy a different part of the binding site of a
biological receptor. Display methods are quite good tools for the identification of compounds, samples, or objects which have different features
to the rest of the set.

Figure 4.18 illustrates the use of the power term, ρ, in Equation (4.3).
The bicyclic amine data set shown in Figure 4.15 was mapped using a
value of two for this term. With $\rho = 2$, both large and small interpoint
distances are equally preserved; this compromise ensures the best overall
mapping of the P-space interpoint distances. Figure 4.18 shows the result
of mapping this same data set using a value of -2 for ρ. This has the effect

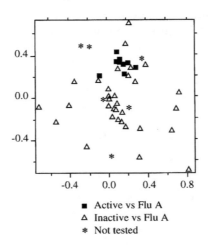

Figure 4.18   NLM, using a power term $\rho = -2$, of the antiviral bicyclic amine
derivatives shown in Figure 4.15 (reproduced from ref. [3] with kind permission of
Springer Science + Business Media).

of preserving the larger interpoint distances at the expense of the smaller ones; the result is to 'collapse' local clusters of points thus emphasizing the similarities between compounds. The effect on the data set has been quite dramatic; the active compounds still cluster together and it can be seen that none of the test set compounds join this cluster. However, one of the test set compounds now lies very close to the cluster of actives and thus becomes a much more interesting synthetic target. Two of the remaining test set compounds are close together (only one need be made) and one of the test set compounds has been separated from the rest of the set. This latter compound may now represent an interesting target to make, as it may be chemically different to the rest of the test set, or may be ignored since it lies a long way from the active cluster. Synthetic feasibility and the judgement of the research team will decide its fate.

Another example of the use of display methods also involves a different type of descriptor data, results from a panel of human testers. In the analysis of natural orange aroma (NOA) samples reported earlier [9] a human testing panel was trained over a period of three months using pure samples of 15 identified components of the NOA samples. A quantitative descriptive analysis (QDA) report form was devised during the course of the training; the QDA form was used to assign a score to a number of different properties of the NOA samples. PCA of the QDA data for the same samples as shown in Figure 4.14 resulted in the explanation of 58 % of the variance in the data set in the first three PCs. A scores plot of these three PC axes is shown in Figure 4.19 where it can be seen that the NOA samples are broadly grouped together into different categories, but the classifications are not as tight as those shown in Figure 4.14.

Figure 4.20 shows a nonlinear map of Fisher-weighted QDA where it can be seen that some of the categories are quite well separated but not as clearly as the NLM from GC–MS data (see Figure 4.16).[4]

Some of the advantages and disadvantages of nonlinear mapping as a multivariate display technique are listed in Table 4.5. Most of these have been discussed already in this section but a couple of points have not.

Since the technique is an unsupervised learning method, it is unlikely that any grouping of objects will happen by chance. Any cluster of points seen on an NLM generally represents a cluster of points in the P-dimensional space. Such groupings may happen by chance although this is much more likely to occur when a supervised learning method, which seeks to find or create patterns in a data set, is employed. The significance

---

[4] Fisher-weighting and variance-weighting are different procedures for weighting variables according to their ability to classify samples (see ref. [17]).

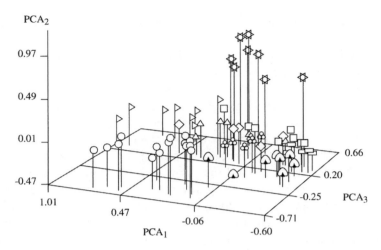

Figure 4.19 Scores plot for a set of NOA samples described by sensory QDA data. The QDA data was autoscaled and variance-weighted (see reference for details). Symbols are the same as those used in Figure 4.14 (reproduced from ref. [9] with permission of Elsevier).

of a group of points found on a nonlinear map, or any other display for that matter, may be assessed by a method called *cluster significance analysis* as discussed in Chapter 5. The fact that the display is dependent on the order of the compounds and changes as compounds are added or

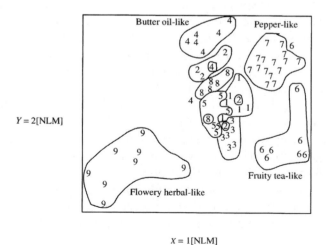

Figure 4.20 NLM of a set of NOA samples described by Fisher-weighted sensory QDA data. Symbols are the same as those used in Figure 4.14 (reproduced from ref. [9] with permission of Elsevier).

Table 4.5 Nonlinear mapping – pros and cons.

Advantage
No assumptions concerning mechanism and may identify different mechanisms
Unsupervised learning so chance effects unlikely
Does not require biological data
Non-linear
Can change the emphasis on the preservation of interpoint distances
Can view multivariate data in two (or three) dimensions

Disadvantage
Unknown non-linear combination of variables
Cannot plot a point directly on the map
Display may change dramatically as points are added/removed
Cannot relate NLM distances to N-space distances (mapping errors)
Display depends on the order of data entry

removed is a consequence of the minimization of the error function. The calculated map depends on the initial guess for the 2-space points since the minimizer will find the nearest local minimum rather than the global minimum (if one exists). A common way to choose the initial positions of the points in 2-space is to assign them randomly, but a disadvantage of this is that running the NLM routine several times on the same data set may produce several different maps. One approach to overcoming this problem is to use principal component scores as the initial guess for the 2-space positions; a disadvantage of this is that the resultant map may be more 'linear' than is desirable. Since the error function is calculated over a summation of the distance differences, adding or removing points may alter the subsequent display. This can be disconcerting to newcomers to the method, particularly when we are accustomed to display methods which give only one 'answer'.

Finally, most of the examples shown here have been two-dimensional but the addition of an extra dimension can dramatically improve the performance of a display method such as PCA or NLM. The data shown in Table 4.6 is a set of computed descriptors for 26 compounds which are antagonists of the $5HT_3$ receptor. $5HT_3$ antagonists are highly effective at preventing nausea and vomiting during chemotherapy and radiotherapy.

The molecules were originally described by a total of 112 computed properties and therefore, as described in Chapter 3, the starting data set contained a considerable amount of redundancy. Elimination of pairwise correlations reduced this set to 56 and the application of a feature selection routine (see Chapter 7) chose the 9 variables shown in the table

Table 4.6 Calculated molecular descriptors for $5HT_3{}^*$ antagonists based on the parent structure shown below (reproduced from ref. [21] with permission of Elsevier).

Compound number	Activity*	CMR	μZ	HOMO	ALP(3)	FZ(4)	VDWE(4)	FY(6)	FZ(9)	FY(11)
1	1	83.156006	−0.023075	−10.125457	0.165465	0.690959	−0.01822	0.146858	−1.331931	−0.571263
2	1	85.682991	1.133915	−10.373082	0.164176	0.825953	−0.067104	−0.095023	−1.814457	−1.311374
3	1	83.169998	−1.009328	−10.291448	0.166767	0.199967	−0.007455	0.131438	−0.261507	1.610443
4	1	92.446007	−1.100368	−10.269679	0.166118	0.350187	−0.092965	−0.355433	−0.178441	3.939417
5	1	93.903	−0.883712	−10.767652	0.165174	0.323102	−0.09015	0.245022	−0.444854	−0.101309
6	1	90.302002	1.681585	−9.896879	0.165143	0.263393	−0.091509	−0.230816	−0.693684	−2.056473
7	1	86.670998	1.509192	−10.141784	0.16502	0.012006	−0.08891	−0.466338	−0.280089	−1.676208
8	1	90.302002	0.374931	−9.85334	0.165122	0.026151	−0.102998	−0.449422	−0.066987	−1.326878
9	1	87.794006	−2.013182	−10.100966	0.166473	0.240704	−0.071487	0.281943	−0.05055	1.565157
10	1	92.19001	−1.59107	−10.354034	0.165545	−0.403645	−0.237743	−0.702534	−0.465868	0.124084
11	1	87.814011	1.266109	−10.571728	0.160491	0.151386	−0.042395	−0.152652	−0.58176	0.474638
12	1	78.888992	−0.01763	−11.056095	0.164764	−0.31937	−0.017028	−0.365937	0.729973	0.701252
13	1	85.682991	−2.126862	−10.373082	0.165448	1.154631	−0.0777	0.106626	−1.773054	−0.115845
14	1	85.682991	−1.260413	−10.457438	0.165508	0.956636	−0.059637	0.089594	−1.630651	−0.489548
15	1	86.75	0.445967	−10.813911	0.163938	0.051509	−0.013655	−0.101951	0.724446	−0.264842

#	*	CMR	μZ	HOMO					VDWE	ALP
16	1	86.751007	-0.130124	-10.337708	0.163203	0.342944	-0.054695	0.088702	-1.062614	-1.835011
17	1	81.836998	-0.129692	-10.008447	0.165598	0.304008	-0.08273	0.202748	-0.121693	2.285414
18	1	81.808998	-1.471474	-11.050653	0.164786	0.155182	-0.052069	-0.053286	-0.372007	0.229041
19	2	87.701996	-0.567377	-10.759488	0.163277	0.334046	-0.096817	0.083781	-1.089644	-1.575862
20	2	80.582001	-0.56162	-10.446554	0.164764	0.274729	-0.061855	0.053127	-0.56334	-0.273451
21	2	83.063995	-0.992965	-10.841124	0.163602	0.37626	-0.02528	0.068368	-1.002572	-1.531959
22	2	81.836998	0.665133	-10.383967	0.16327	0.280005	-0.054866	-0.004466	-1.071213	-2.130827
23	2	79.928001	-0.683407	-10.109129	0.166261	0.041216	-0.01325	-0.019164	0.003597	2.001035
24	2	89.233002	-0.955246	-10.55268	0.163597	0.611623	-0.217868	0.116554	-1.201934	-1.457719
25	2	85.49601	-0.464845	-10.503699	0.164757	0.153127	-0.084228	-0.013756	-0.481711	-0.30028
26	2	87.459	-1.887398	-10.830238	0.163055	0.003264	-0.042777	0.031053	-3.251751	-1.093677

* 1 = inactive; 2 = active

CMR calculated molar refractivity

μZ Z component of the dipole moment

HOMO energy of the highest occupied molecular orbital

FZ(N°) and FY(N°) Z and Y components of the electric field at specified (N°) grid points

VDWE(N°) the Van der Waal's energy of the interaction of a carbon atom at a specified (N°) grid point

ALP(N°) the self atom polarizability of the specified atom (N°)

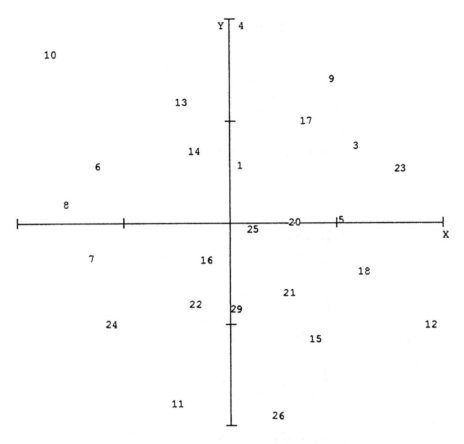

Figure 4.21   Two-dimensional NLM from the data in Table 4.6 (map created using the SciFit package – www.scimetrics.com).

as being important for the prediction of activity. A two-dimensional NLM was calculated from these descriptors and gave some clustering of the active compounds (19 to 26) as shown in Figure 4.21.

The clustering is not very convincing, however, and there is little real separation between active and inactive compounds. A two-dimensional scores plot on the first two PC's gave an even more confused picture with some clustering of the active compounds together but with inactives mixed in as well. A three-dimensional NLM, however, gives an almost perfect separation of the active compounds as shown in Figure 4.22.

All of the active compounds, except 23, are collected together in a single region of the 3D map bounded by the inactives 11, 13, 14 and 15. Clearly, the compression of information from a 9 variable space to two

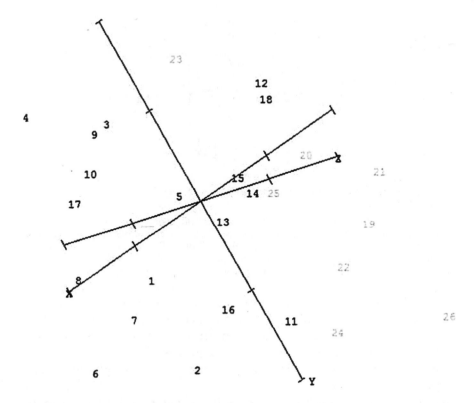

Figure 4.22   Three-dimensional NLM from the data in Table 4.6 (map created using the SciFit package – www.scimetrics.com).

dimensions loses too much information compared with the corresponding three dimensional plot.

4.3.2   Self-organizing Map

A self-organizing map (SOM) or Kohonen map [18] is an artificial neural network (ANN) architecture used for the display of multivariate data. ANN are described in detail in Chapter 9 so suffice to say here that the 'heart' of these networks is an artificial neuron, designed to mimic to some extent biological neurons, and that these artificial neurons can be connected together in a variety of different patterns or architectures. The neurons are connected to one another via a set of weights, called connection weights, and training of the ANN involves alteration of these weights by a training algorithm until some training target is met.

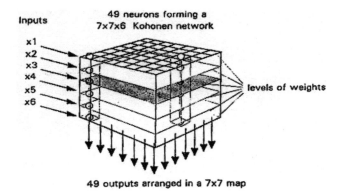

Figure 4.23 Block representation of a 7×7 Kohonen network with 6 input variables. Each layer of the block depicts the connection weights between a particular input variable and all of the neurons in the 7×7 plot. The shaded layer shows the weights for input variable 3 with the particular weights for neurons (1,1) and (6,5) picked out as shaded circles (reproduced from ref. [19] with permission).

A SOM is made up from a two-dimensional array of neurons, each one being connected, via a connection weight, to each of the inputs (variables in the data set). The set of weights and neurons can be represented by a three-dimensional diagram as shown in Figure 4.23.

In the figure a single layer in the 3-D 'stack' represents all the weights associated with a particular input variable. Training a SOM consists of two parts, competitive learning and self-organization. Initially, as with most network training procedures, the connection weights are set to random values. Each pattern (object, case, compound) in the training set may be considered to be a vector, X, consisting of p values x_i (where there are p variables in the set); each neuron j in the map is characterized by a weight vector, W_j, consisting of p weights w_{ij}. Euclidean distances, d_j, are calculated between each X and each weight vector W_j by Equation (4.4):

$$d_j = \left(\sum_{i=1}^{m} (x_i - w_{ij})^2 \right)^{1/2} \tag{4.4}$$

The neuron with the weight vector W_j closest to the input pattern X is said to be the winning neuron, j^*, and it is updated so that its weight vector, W_j^*, is even closer to the input vector X:

$$w_{ij}^*(t + 1) = w_{ij^*}(t) + \alpha(t)[x_i - w_{ij}^*(t)]$$
$$0 < \alpha < 1 \tag{4.5}$$

The terms t and α in Equation (4.5) are time and learning rate, respectively, (t and $t+1$ are successive instants of time) and after each updating the time variable is incremented whereas the learning rate, α, is decreased. This process, which is called a step, is repeated for each of the patterns in the training set; a learning epoch consists of as many steps as there are patterns in the training set.

The competitive learning phase of Kohonen mapping takes no account of the topological relationships between the neurons in the plot, these are updated in an isolated fashion. Self-organization is the second phase of the training process and this is achieved by defining a set of neighbouring neurons, N_{j*}, as the set of neurons which are topologically close to the 'winning' neuron j^*. The learning algorithm shown in Equation (4.5) may be modified so that the neighbouring neurons are updated as well as the j^* neuron:

$$
\begin{aligned}
w_{ij}(t+1) &= w_{ij}(t) + \alpha(t)\gamma(t)[x_i - w_{ij}(t)] \\
\gamma(t) &= 1 \qquad \forall_j \in N_{j*}(t) \\
\gamma(t) &= 0 \qquad \forall_j \notin N_{j*}(t)
\end{aligned}
\tag{4.6}
$$

The neighbouring neurons are specified by the parameter γ and at first the area of this set is wide but, as training proceeds, the radius of this area is decreased (like the learning rate α) as the time variable t is incremented. The result of this combination of competitive learning and self-organization is to produce a two-dimensional plot in which the data points are arranged according to their similarities in the high dimensional space defined by the variables in the data set. The similarity between SOM mapping and NLM or principal component scores plots is evident.

Table 4.7 shows descriptor data and disconnection mechanism for a set of 32 carbonyl containing compounds. The four different disconnection mechanisms, Aldol-, Claisen-, Michael- and enamine, indicate the way that these compounds react in a retrosynthetic analysis. Differences in reaction mechanism are due to different substitution patterns (not shown) on the compounds.

Bienfait [20] created a SOM for this data set as shown in Figure 4.24. The training process was run for a total of 20 epochs and at the end of this time the resultant map was clearly split into four distinct areas as shown in the bottom right of the figure.

The figure shows four of the stages of this training process, including the initial and final plots, where it can be seen that the four different classes of disconnection have been separated in the display. Although

Table 4.7 Carbonyl compounds, molecular descriptors and disconnection mechanism (reproduced from ref. [20] with permission of the American Chemical Society).

Disconnection	Index	C_ε	C_δ	C_γ	C_β	C_α	$C_{\alpha'}$	$C_{\beta'}$
Aldol-type	1	0	0	0.1	0.36	0.08	0.02	0
	2	0	0	0.1	0.38	0.08	0.1	0
	3	0	0.1	0.08	0.36	0.06	0.02	0
	4	0	0.1	0.08	0.38	0.06	0.1	0
Claisen-type	5	0	0	0.02	0.3	0.08	0.18	0.1
	6	0	0	0.02	0.3	0.08	0.18	0.08
	7	0	0	0.1	0.3	0.06	0.18	0.1
	8	0	0	0.1	0.3	0.06	0.18	0.08
Michael-type	9	0.02	0.3	0.9	0.08	0.08	0.02	0
	10	0.02	0.3	0.9	0.06	0.08	0.02	0
	11	0.02	0.3	0.9	0.08	0.06	0.02	0
	12	0.02	0.3	0.9	0.04	0.08	0.02	0
	13	0.02	0.3	0.9	0.06	0.06	0.02	0
	14	0.02	0.3	0.9	0.04	0.06	0.02	0
enamine-type	15	0.02	0.3	0.08	0.08	0.08	0.02	0
	16	0.02	0.3	0.08	0.08	0.06	0.02	0
	17	0.02	0.3	0.08	0.06	0.08	0.02	0
	18	0.02	0.3	0.06	0.08	0.08	0.02	0
	19	0.02	0.3	0.08	0.06	0.06	0.02	0
	20	0.02	0.3	0.08	0.04	0.08	0.02	0
	21	0.02	0.3	0.06	0.06	0.08	0.02	0
	22	0.02	0.3	0.04	0.08	0.08	0.02	0
	23	0.02	0.3	0.06	0.08	0.06	0.02	0
	24	0.02	0.3	0.06	0.04	0.08	0.02	0
	25	0.02	0.3	0.06	0.06	0.06	0.02	0
	26	0.02	0.3	0.04	0.06	0.08	0.02	0
	27	0.02	0.3	0.08	0.04	0.06	0.02	0
	28	0.02	0.3	0.04	0.08	0.06	0.02	0
	29	0.02	0.3	0.06	0.04	0.06	0.02	0
	30	0.02	0.3	0.04	0.06	0.06	0.02	0
	31	0.02	0.3	0.04	0.04	0.08	0.02	0
	32	0.02	0.3	0.04	0.04	0.06	0.02	0

the classes have been separated, however, it can also be seen that the final map does not display all of the compounds in the set. The four compounds which undergo Claisen-type disconnection, for example, are represented by just a single cell (neuron) on the plot while the six Michael-type compounds are shown in just two cells.

Principal component analysis of this data set resulted in two principal components which explained ~76 % of the variance in the data. A scores

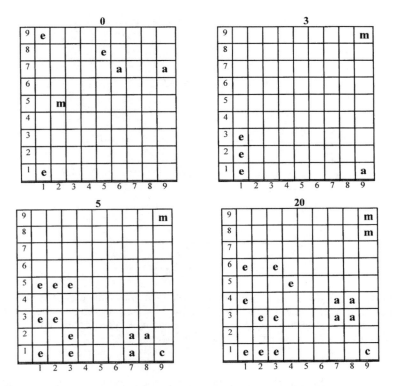

Figure 4.24 Four stages in the training of a Kohonen self-organizing map for the data shown in Table 4.7. Numbers above each plot indicate the number of training epochs and the letters a, c, m and e represent Aldol, Claisen, Michael and enamine-type disconnections respectively (reproduced from ref. [20] with permission of the American Chemical Society).

plot from this PCA is shown in Figure 4.25 where it can be seen that the Aldol- and Claisen-type disconnections are clearly separated from the other two classes, although these fall into two groups.

The Michael- and enamine-type disconnections fall into two mixed clusters. A nonlinear map of the same data is shown in Figure 4.26. Here it can be seen that compounds which follow all four disconnection mechanisms are clearly separated although, again, each class is split into two separate groups.

Both the PC scores plot and the NLM appear to be giving a better view of the data since most of the compounds are visible, compared with the display of the SOM. The latter, however, has a limited resolution of 9 × 9 pixels since this was the size chosen for the map. Increasing the number of neurons in the display may well separate the overlapping

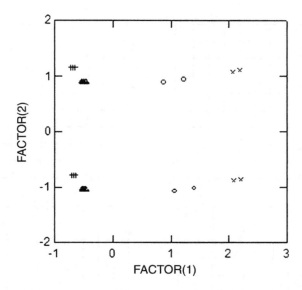

Figure 4.25 Scores plot on the first two principal component axes for the compounds shown in Table 4.7, points are shown as a circle for Aldol- a cross for Claisen- a plus for Michael- and a triangle for enamine-type disconnections (reproduced from ref. [21] with permission of Elsevier).

compounds although this is likely to be at the expense of longer training times since there are more connection weights to compute. The SOM or Kohonen map is claimed to be one of the most widely applied neural network algorithms [22] and there are certainly many examples of it's use in a wide variety of application areas.

4.4 FACES, FLOWERPLOTS AND FRIENDS

The previous display methods have all been examples of dimension reduction; that is to say they are means by which a high dimensional data set can be shown in a lower dimensional plot such as 2- or 3-D. This has been achieved by creating new variables which are linear or non-linear combinations of the original variables. The advantage of this is that all of the information in the original variables is used but the disadvantage is that, in combining the variables, the 'compression' may obscure some useful information. To avoid this problem what is needed is some means by which all of the variables can be simultaneously displayed but without creating combinations of them. Fortunately, there are some

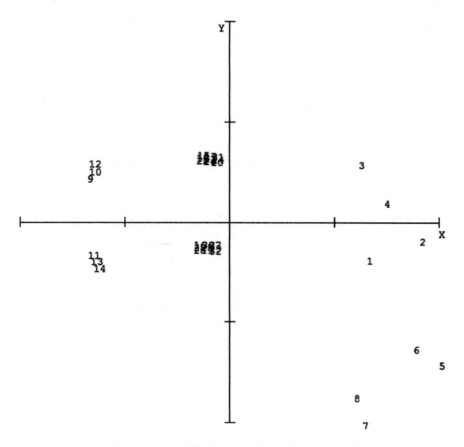

Figure 4.26 Non-linear map of the compound set shown in Table 4.7, points are numbered as in the table that is 1-4 Aldol-, 5-8 Claisen- and so on. Display generated using the SciFit package – www.scimetrics.com (reproduced from ref. [21] with permission of Elsevier).

ingenious techniques which can be used to display high dimensional data in lower dimensions and these make use of our natural ability to recognize patterns.

Perhaps the most obvious example of human ability to recognize patterns is the way that we can identify faces. This was exploited by Herman Chernoff when he devised the method that bears his name, Chernoff faces [23]. In this technique the facial characteristics of a cartoon face, such as size of ears, shape of mouth, slant of eyebrows, size of nose and so on, are assigned to each of the variables in the set. The result is a single face for each case in the set and similar cases can be rapidly identified as they

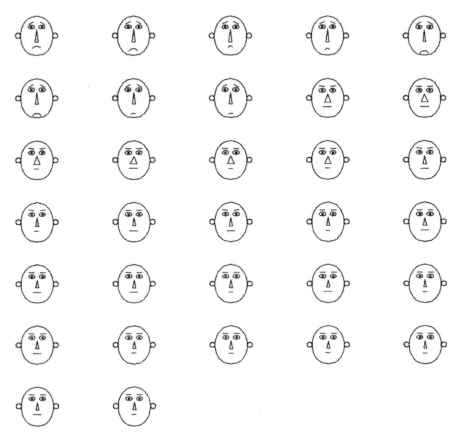

Figure 4.27   Chernoff faces display of the data in Table 4.7. The data was standard-ized (autoscaled, see Section 3.3) and the cases are shown in the order 1–5, left to right top row, 6–10 l-r second row and so on. Display generated using the statistics package Systat (www.systat.com).

have similar faces. Figure 4.27 shows the data in Table 4.7 displayed as Chernoff faces.

The Aldol- and Claisen-type disconnections are quite clearly distin-guished from the other compounds although they are not so easily sep-arated from one another. The Michael-type disconnections have char-acteristic large noses and the enamines are shown up as having straight mouths, eyebrows and smaller noses than the Michael-type. This type of display can work quite well although it becomes difficult to use for reasonably large numbers of samples.

Chernoff faces are a particular example of a type of plot known as icon plots. Icon plots use a geometric shape or object, the icon, which has sufficient features whose characteristics can be altered so that the variables can be displayed. A popular icon plot is the star plot as shown in Figure 4.28.

The star plot operates by assigning individual 'rays' of a symbolic representation of a star to each of the variables. The length of a ray for an individual case represents the magnitude of that variable for the case relative to the maximum magnitude of the variable across all the cases. The star plot in Figure 4.28 quite clearly shows all four classes of compounds and easily separates the Aldol- and Claisen-type disconnections. In this respect it has performed better than the Chernoff faces but this is really demonstrating a common property of all data display methods, whether dimension reducing or dimension preserving, and that is that there is no 'best' technique. All have their advantages and disadvantages and, depending on the data set, some will work better or worse than others. A final type of icon plot which is quite popular is the flower plot. A flower plot is constructed by assigning variables to 'petals' arranged on a circle and, as in the case of other icon plots, each case has a single flower plot symbol. Negative values of variables are shown as the petals going inside the circle and positive going out from the circle. Figure 4.29 shows the now familiar data from Table 4.7 displayed as flower plots where, once again, all four types of compounds can be distinguished.

4.5 SUMMARY

Multivariate display methods are very useful techniques for the inspection of high-dimensional data sets. They allow us to examine the relationships between points (compounds, samples, etc.) in both training and test sets, and between descriptor variables. Dimension reduction can be achieved using linear and non-linear methods, both with advantages and disadvantages, and this has proved useful in numerous scientific applications. The linear approach (PCA) forms the basis of a variety of multivariate techniques as described later in this book. Other techniques for the display of multidimensional data in fewer dimensions, but without recourse to combinations of the original data, can be useful for moderately sized data sets. Finally, it is not possible to say in advance which, if any, is the best approach to use.

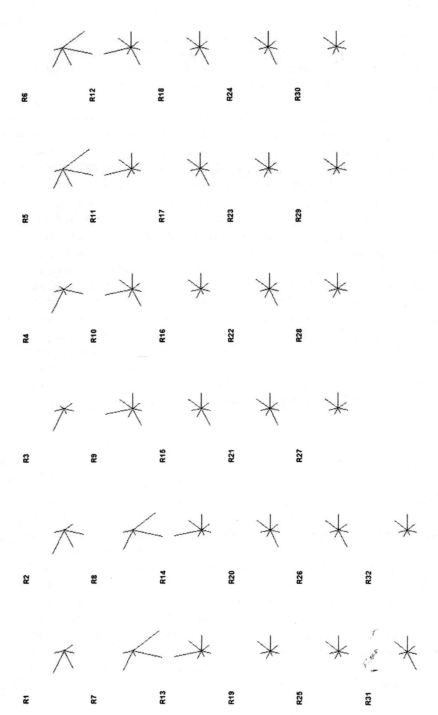

Figure 4.28 Star plot display of the data in Table 4.7. The data was standardized and case labels are shown as R1, R2 and so on. The plot was created using the SciFit package – www.scimetrics.com.

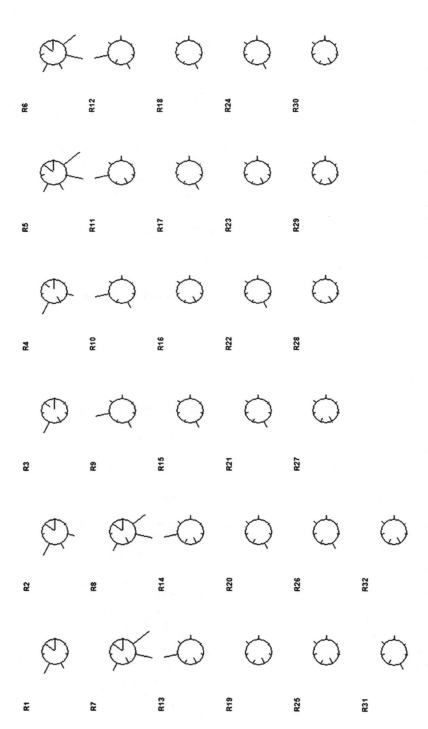

Figure 4.29 Flower plot display of the data in Table 4.7. The data was standardized and case labels are shown as R1, R2 and so on. The plot was created using the SciFit package – www.scimetrics.com.

In this chapter the following points were covered:

1. how the selection of appropriate variables can give the most useful 'view' of a data set;
2. the need for the inclusion of more variables in displays of multivariate data;
3. the way that principal components analysis works and the meaning of scores and loadings;
4. the interpretation of scores plots and loadings plots;
5. how to rotate principal components and the effects of such rotations;
6. non-linear methods for reducing the dimensionality of a data set in order to display it in lower dimensions;
7. how artificial neural networks can be used to produce low dimensional plots of a multivariate data set;
8. what icon plots are and how they can be used to display multiple variables without data compression.

REFERENCES

[1] Lewi, P.J. (1986). *European Journal of Medicinal Chemistry*, **21**, 155–62.
[2] Seal, H. (1968). *Multivariate Analysis for Biologists*. Methuen, London.
[3] Hudson, B.D., Livingstone, D.J., and Rahr, E. (1989). *Journal of Computer-aided Molecular Design*, **3**, 55–65.
[4] Dizy, M., Martin-Alvarez, P.J., Cabezudo, M.D., and Polo, M.C. (1992). *Journal of the Science of Food and Agriculture*, **60**, 47–53.
[5] Ghauri, F.Y., Blackledge, C.A., Glen, R.C., *et al.* (1992). *Biochemical Pharmacology*, **44**, 1935–46.
[6] Van de Waterbeemd, H., El Tayar, N., Carrupt, P.-A., and Testa, B. (1989). *Journal of Computer-aided Molecular Design*, **3**, 111–32.
[7] Jackson, J.E. (1991). *A User's Guide to Principal Components*, pp. 155–72. John Wiley & Sons, Inc., New York.
[8] Livingstone, D.J. (1991). Pattern recognition methods in rational drug design. In *Molecular Design and Modelling: Concepts and Applications, Part B, Methods in Enzymology*, Vol. 203 (ed. J.J. Langone), pp. 613–38. Academic Press, San Diego.
[9] Lin, J.C.C., Nagy, S., and Klim, M. (1993). *Food Chemistry*, **47**, 235–45.
[10] Gabriel, K.R. (1971). *Biometrika*, **58**, 453–67.
[11] Shepard, R.N. (1962). *Psychometrika*, **27**, 125–39, 219–46.
[12] Kruskal, J.B. (1964). *Psychometrika*, **29**, 1–27.
[13] Sammon, J.W. (1969). *IEEE Transactions on Computers*, **C-18**, 401–9.
[14] Kowalski, B.R. and Bender, C.F. (1973). *Journal of the American Chemical Society*, **95**, 686–93.

[15] Digby, P.G.N. and Kempton, R.A. (1987). *Multivariate Analysis of Ecological Communities*, pp. 19–22. Chapman & Hall, London.
[16] Clare, B.W. (1990). *Journal of Medicinal Chemistry*, **33**, 687–702.
[17] Varmuza, K. (1980). *Pattern Recognition in Chemistry*, pp. 106–9. Springer-Verlag, New York.
[18] Kohonen, T. (1990). *Proceedings of the IEEE*, **78**, 1464–80.
[19] Zupan, J. (1994). *Acta Chimica Slovenica*, **41**, 327–52.
[20] Bienfait, B. (1994). *Journal of Chemical Information and Computer Science*, **34**, 890–8.
[21] Livingstone, D.J. (1996). Multivariate data display using neural networks. In *Neural Networks in QSAR and Drug Design* (ed. J. Devillers), pp. 157–76. Academic Press, London.
[22] Oja, E. and Kaski, S. (eds) (1999). *Kohonen Maps*, Elsevier, Amsterdam.
[23] Chernoff, H. (1973). *Journal of American Statistical Association*, **68**, 361–8.

5

Unsupervised Learning

Points covered in this chapter
- k-nearest neighbours
- Factor analysis
- Cluster analysis
- Cluster significance analysis

5.1 INTRODUCTION

The division of topics into chapters is to some extent an arbitrary device to produce manageable portions of text and, in the case of this book, to group together more or less associated techniques. The common theme underlying the methods described in this chapter is that the property that we wish to predict or explain, a biological activity, chemical property, or performance characteristic of a sample, is not used in the analytical method. Oddly enough, one of the techniques described here (nearest-neighbours) does require knowledge of a dependent variable in order to operate, but that variable is not directly involved in the analysis. The display methods described in Chapter 4 are also unsupervised learning techniques, and could have been included in this section, but I felt that display is such a fundamental procedure that it deserved a chapter of its own. Cluster analysis, described in Section 5.4, may also be thought of as a display method since it produces a visual representation of the relationships between samples or parameters. Thus, the division between display methods and unsupervised learning techniques is mostly artificial.

A Practical Guide to Scientific Data Analysis David Livingstone
© 2009 John Wiley & Sons, Ltd

5.2 NEAREST-NEIGHBOUR METHODS

A number of different methods may be described as looking for nearest neighbours, e.g. cluster analysis (see Section 5.4), but in this book the term is applied to just one approach, k-nearest-neighbour. The starting point for the k-nearest-neighbour technique (KNN) is the calculation of a distance matrix as required for non-linear mapping. Various distance measures may be used to express the similarity between compounds but the Euclidean distance, as defined in Equation (4.2) (reproduced below), is probably most common:

$$ d_{ij} = \sqrt{\left(\sum_{k=1,P} (d_{i,k} - d_{j,k})^2 \right)} \qquad (5.1) $$

where d_{ij} is the distance between points i and j in P-dimensional space. A distance matrix is a square matrix with as many rows and columns as the number of rows in the starting data matrix. Table 5.1 shows a sample distance matrix from a data set containing ten samples. The diagonal of this matrix consists of zeroes since this represents the distance of each point from itself. The bottom half of the matrix gives the distance, at the intersection of a row and column, between the samples represented by that row and column; the matrix is symmetrical, i.e. distance B → A = distance A → B, so the top half of the matrix is not shown here. An everyday example of a distance matrix is the mileage chart, which can be found in most road atlases, for distances between cities.

The classification of any unknown sample in the distance matrix may be made by consideration of the classification of its nearest neighbour.

Table 5.1 Distance matrix for ten samples.

	A	B	C	D	E	F	G	H	I	J
A	0									
B	1.0	0								
C	2.6	2.5	0							
D	2.8	2.6	1.3	0						
E	3.2	2.2	2.8	2.1	0					
F	3.4	2.4	3.1	3.0	1.3	0				
G	3.7	3.4	4.1	3.0	1.3	1.3	0			
H	6.2	5.3	4.3	3.0	3.0	3.2	2.9	0		
I	9.8	9.7	4.0	3.7	6.2	7.5	6.2	3.5	0	
J	10.0	9.9	4.4	4.0	6.3	7.6	6.4	3.6	1.2	0
	A	B	C	D	E	F	G	H	I	J

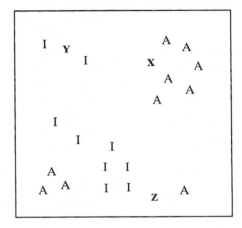

Figure 5.1 Two-dimensional representation of the KNN technique; training set compounds are represented as A for active and I for inactive, test set compounds as X, Y, and Z.

This involves scanning the row and column representing that sample to identify the smallest distance to other samples. Having identified the distance (or distances) it is assumed that the classification of the unknown will be the same as that of the nearest neighbour, in other words samples that are similar in terms of the property space from which the distance matrix was derived will behave in a similar fashion. This mimics the 'common-sense' reasoning that is customarily applied to the interpretation of simple two-dimensional plots, the difference being that here the process is applied in P-dimensions. Figure 5.1 shows a two-dimensional representation of this process.

The training set compounds are shown marked as A and I for active and inactive; the unknown, test set, compounds are indicated as X, Y, and Z. The nearest neighbour to compound X is active and that to compound Y is inactive, and this is how these two would be classified on the basis of one nearest neighbour. Classification for compound Z is more difficult as its two apparently equidistant neighbours have different activities. Although one of these neighbours may be slightly closer when the values in the distance matrix are examined, it is clear that this represents an ambiguous prediction. With the exception of one close active compound the remaining neighbours of compound Z are inactive and the common-sense prediction would be for Z to be inactive. This is the meaning of the term k in k-nearest-neighbour, k refers to the number of neighbours that will be used for prediction. The choice of a value for k will be determined by the training set; this is achieved by comparing the predictive performance of different values of k for the training set compounds. The value of k can be any number, including 1, but is usually chosen to be an odd number so that the algorithm may make an unambiguous decision. Figure 5.1 also illustrates a quite common situation in the analysis of multivariate data sets; the two activity classes are not linearly separable i.e. it is not possible to draw a single straight line that will divide up the space into two regions containing only active or inactive compounds. Some analytical methods operate by the construction of a hyperplane, the multivariate analogue of a straight line, between classes of compounds (see Chapter 7). In the case of a data set such as this, the KNN method will have superior predictive ability.

Nearest-neighbour methods are also able to make multi-category predictions of activity; training set samples can be ranked into any number of classifications, but it is important to maintain the balance between the number of classes and the number of members within a class. Ideally, each class should contain about the same number of members although in some situations (such as where the property is definitely YES/NO) this may not be possible to achieve. The reason for maintaining similar numbers of members in each class is so that a 'random' prediction of membership for any class is not significantly greater than that for the other classes. This raises the question of how to judge the prediction success of a nearest-neighbour method. When a training set is split into two classes, there is a 50 % chance of making a successful prediction for any compound, given equal class membership. A success rate of 80 % for the training set may sound impressive but is, in fact, little over half as good again as would be expected from purely random guesses. Where classes do differ significantly in size it is possible to change the random expectation in order to judge success. For example, if class 1 contains twice as many members as class 2, the random expectations are ~66 and ~33 % respectively. Another aspect of the balance between the number of classes and the size of class membership concerns the dimensionality of the data set. When the number of dimensions in a data set is close to or greater than the number of samples in the set, it is possible to discover linear separation between classes by chance. The risk of this happening is obviously greatest for supervised learning methods (which are intended to find linear separations) but may also happen with unsupervised techniques. There are no 'rules' concerning the relationship between dimensions and data points for unsupervised learning methods but, as with most forms of analysis, it is desirable to keep the problem as simple as possible.

Table 5.2 Nearest-neighbour classification for NMR (12 features) data set (reproduced from ref. [1] with permission of the American Chemical Society).

	Correct/Total	
	Training set	Test set
1–nearest-neighbour		
Class 1	60/66	60/66
Class 2	60/66	60/66
Class 3	64/66	64/66
Total	184/198 = 93 %	184/198 = 93 %
Learning machine		
Class 1	56/66	54/66
Class 2	12/66	12/66
Class 3	24/66	23/66
Total	92/198 = 46 %	89/198 = 45 %

Now for some examples of the application of nearest-neighbour methods to chemical problems. An early example involved the classification of compounds described by calculated NMR spectra [1]. The data set consisted of 198 compounds divided into three classes (66 each) of molecules containing $CH_3CH_2CH_2$, CH_3CH_2CH, or CH_3CHCH. The NMR spectra were preprocessed (see reference for details) to give 12 features describing each compound and the data set was split in half to give training and test sets. Table 5.2 shows the results for training set and test set predictions using 1–nearest neighbour.

Since there are three classes with equal class membership, the random expectation would be 33 % correct, and thus it appears that the nearest-neighbour technique has performed very well for this set. Also shown in Table 5.2 are the results for a method called the linear learning machine (see Section 7.2.1) which has performed quite poorly with a success rate only slightly above that expected by chance, suggesting that the data is not linearly separable. An application of KNN in medicinal chemistry was reported by Chu and co-workers [2]. In this case the objective was to predict the antineoplastic activity of a test set of 24 compounds in a mouse brain tumour system. The training set consisted of 138 structurally diverse compounds which had been tested in the tumour screen. The compounds, test set, and training set, were described by a variety of sub-structural descriptors giving a total of 421 parameters in all.

Table 5.3 Comparison of predicted and observed antineoplastic activities (reproduced from ref. [2] with permission of the American Chemical Society).

	Non-active	False negative	Active	False positive
KNN	14	1	6	3
Experimental	17		7	

Various procedures were adopted (see reference for details) to reduce this to smaller sized sets and KNN was employed to make predictions using the different data sets. The KNN predictions were averaged over these data sets to give an overall success rate of 83 %. A comparison of the predictions with the experimental results is shown in Table 5.3.

Scarminio and colleagues [3] reported a comparison of the use of several pattern recognition methods in the analysis of mineral water samples characterized by the concentration of 18 elements, determined by atomic absorption and emission spectrometry. The result of the application of cluster analysis and SIMCA to this data set is discussed elsewhere (Sections 5.4 and 7.2.2); KNN results are shown in Table 5.4. The performance of KNN in this example is really quite impressive; for two regions, the training set samples are completely correctly classified up to five nearest-neighbours and the overall success rate is 95 % or better (~25 % success rate for the random expectation). A test set of seven samples was analysed in the same way and KNN was found to classify all the samples correctly, considering up to nine nearest-neighbours.

An example of the use of KNN in chemistry was reported by Goux and Weber [4]. This study involved a set of 99 saccharide residues, occurring as a monosaccharide or as a component of a larger structure, described

Table 5.4 KNN classification results for water samples, collected from four regions, described by the concentration of four elements (Ca, K, Na, and Si) (reproduced from ref. [3] with permission of Energia Nuclear & Agricultura).

Region	Number of samples	Number of points incorrectly classified			
		1-NN	3-NN	5-NN	7-NN
Serra Negra	46	2	3	3	2
Lindoya	24	0	0	0	1
Sao Jorge	7	1	2	1	1
Valinhos	39	0	0	0	0
	Correct (%)	97.3	95.5	96.4	96.4

Table 5.5 Nearest-neighbour classification of glycosides (reproduced from ref. [4] with permission of Elsevier).

Data set	Number misassigned (% correct)
UAR-19 (full set)	1 (99)
DP-12	0 (100)
DP-9	8 (90)
SP-6/1	0 (100)
SP-6/4	1 (99)
SP-6/5	3 (96)

by 19 experimentally determined NMR parameters. The NMR measurements included a coupling constant and both proton and carbon chemical shifts. The aim of the work was to see if the NMR data could be used to classify residues in terms of residue type and site of glycoside substitution to neighbouring residues. A further aim was the identification of important NMR parameters, in terms of their ability to characterize the residues. To this end, a number of subsets of the NMR parameters were created and the performance of KNN predictions, in this case 1-NN, were assessed. Table 5.5 illustrates the results of this 1-NN classification for the full dataset and five subsets. For the full dataset of 19 variables, one residue is misassigned, and for two subsets this misclassification is eliminated. Two of the other subsets give a poorer classification rate, demonstrating that not only can the NMR data set be used to classify the residues but also that the important parameters can be recognized.

5.3 FACTOR ANALYSIS

Principal components analysis (PCA), as described in Chapter 4, is an unsupervised learning method which aims to identify principal components, combinations of variables, which 'best' characterize a data set. Best here means in terms of the information content of the components (variance) and that they are orthogonal to one another. Each principal component (PC) is a linear combination of the original variables as shown in Equation (4.1) and repeated below

$$PC_q = a_{q,1}v_1 + a_{q,2}v_2 + \ldots a_{q,P}v_P \qquad (5.2)$$

The principal components do not (necessarily) have any physical meaning, they are simply mathematical constructs calculated so as to comply

with the conditions of PCA of the explanation of variance and orthogonality. Thus, PCA is not based on any statistical model. Factor analysis
(FA), on the other hand, is based on a statistical model which holds that
any given data set is based on a number of factors. The factors themselves
are of two types; common factors and unique factors. Each variable in
a data set is composed of a mixture of the common factors and a single
unique factor associated with that variable. Thus, for any variable X_i, in
a P-dimensional data set we can write

$$X_i = a_{i,1} F_1 + a_{i,2} F_2 + \ldots \ldots a_{i,p} F_p + E_i \qquad (5.3)$$

where each $a_{i,j}$ is the loading of variable X_i on factor F_j, and E_i is the
residual variance specific to variable X_i. The residual variance is also
called a unique factor associated with that variable, the common factors
being F_1 to F_p which are associated with all variables, hence the term
common factors. The similarity with PCA can be seen by comparison of
Equations (5.3) and (5.2). Indeed, PCA and FA are often confused with
one another and since the starting point for a FA can be a PCA this is
perhaps not surprising.

The two methods, although related, are different. In the description
of Equation (5.3) the loadings, $a_{i,j}$, were described as the loadings of
variables X_i on factor F_j so as to point out the similarity to PCA. Expression of this equation in the (hopefully) more familiar terms of PCA gives
these loadings as the loadings of each of the common factors, F_1 to F_p,
onto variable X_i. In other words, PCA identifies principal components
which are linear combinations of the starting variables; FA expresses
each of the starting variables as a linear combination of common factors. PCA seeks to explain all of the variance in a data set; FA seeks
to factor (hence the name) the variance in a data set into common and
unique factors. The unique factors are normally discarded, since it is
usually assumed that they represent some 'noise' such as experimental
error, and thus FA will reduce the variance of a data set by (it is hoped)
removing irrelevant information. Since the unique factors are removed,
the remaining common factors all contain variance from at least two, if
not more, variables. Common factors are explaining covariance and thus
FA is a method which describes covariance, whereas PCA preserves and
describes variance. Like PCs, the factors are orthogonal to one another
and various rotations (like varimax, see Section 4.2) can be applied in
order to simplify them. One of the advantages claimed for FA is that it
is based on a 'proper' statistical model, unlike PCA, and that by discard-
ing unique factors the data set is 'cleaned up' in terms of information

content. FA, however, relies on a number of assumptions and these are equally claimed as disadvantages to the technique. Readers interested in further discussion of FA and PCA should consult Chatfield and Collins [5], Malinowski [6], or Jackson [7].

What about applications of FA? An interesting example was reported by Li-Chan and co-workers [8] who investigated the quality of hand-deboned and mechanically deboned samples of meat and fish. The samples were characterized by physicochemical properties such as pH, fat, and moisture content, and by functional properties such as gel strength and % cookloss (in terms of weight). Factor analysis of the overall data set of 15 variables for 230 samples extracted three factors which described 70 % of the data variance. The factor loadings are shown in Table 5.6 where it can be seen that factor 1 includes the hydrophobic/hydrophilic properties of the salt-extractable proteins, factor 2 describes total and salt-extractable proteins and mince pH, and factor 3 is associated mainly with moisture and fat. The factor loadings may be plotted against one another as for PC loadings (Figure 4.12) in order to show the relationships between variables. Rotated factor loadings from Table 5.6 are shown in Figure 5.2 in which various groupings of associated variables may be seen. For example, total and salt-extractable protein are associated, as are solubility, dispersibility, and emulsifying capacity. Factor scores may be calculated for the samples and plotted on the factor axes. Figure 5.3 shows factor scores for factor 3 versus factor 2 where it

Table 5.6   Sorted rotated factor loadings (pattern) from factor analysis of meat and fish data* (reproduced from ref. [8] with permission of Wiley-Blackwell).

	Factor 1	Factor 2	Factor 3
Dispersibility	−0.959	0.000	0.000
Solubility	−0.939	0.000	0.000
ANS	0.864	0.000	0.255
Gel-M	0.862	0.000	0.000
Gel-E	0.853	0.288	0.000
EC	−0.848	0.000	0.000
CPA	0.844	0.000	0.000
FBC	−0.604	0.293	0.000
Cookloss	0.529	0.000	−0.457
Protein	0.000	0.916	0.000
Mince-pH	0.000	−0.852	0.285
S.E.P.	0.000	0.821	0.000
SH	0.000	0.617	0.000
Moisture	0.000	0.000	0.966
Fat	0.000	−0.382	−0.891

*Loadings less than 0.25 have been replaced by zero.

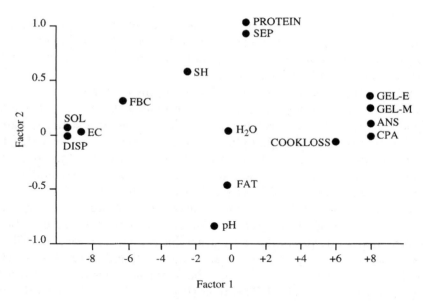

Figure 5.2 Loadings plot on the first two factors for 15 variables used to describe samples of meat and fish. SOL, solubility; EC, emulsifying capacity; DISP, dispersibility; FBC, fat binding capacity; SH, sulphydryl content; PROTEIN, protein content; SEP, salt-extractable protein; H$_2$O, moisture content; pH, pH of a mince suspension; FAT, crude fat content; COOKLOSS, percentage weight lost after cooking; GEL-M, gel strength of mince; GEL-E, gel strength of extract; ANS and CPA, protein surface hydrophobicity using aromatic (ANS) or aliphatic (CPA) fluorescent probe (reproduced from ref. [8], with permission from Wiley-Blackwell).

can be seen that the fish samples are quite clearly distinguished from the meat samples which, in turn, fall into two groups, hand deboned and mechanically deboned.

Takagi and co-workers [9] applied FA to gas chromatography retention data for 190 solutes measured using 21 different stationary phases. Three factors were found to be sufficient to explain about 98 % of the variance of the retention data; physicochemical meanings to these factors were ascribed as shown below:

Factor 1: size
Factor 2: polarity
Factor 3: hydrogen-bonding tendency

As is usually the case with PCA, the attribution of any physical meaning to factors is not straightforward, particularly for the 'later' factors (smaller eigenvalues, less variance explained) from an analysis; this was the case for the third factor. Factor loadings for the three factors

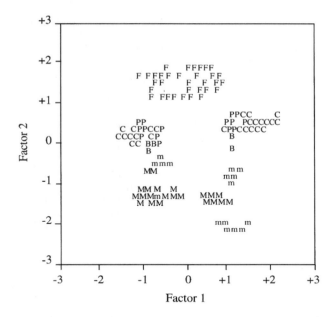

Figure 5.3 Plot of factor scores for meat and fish samples. C, B, P – hand-deboned chicken, beef, and pork; M, m, mechanically deboned pork and chicken; F, cod fish; f, cod fish in presence of cryoprotectants (reproduced from ref. [8] with permission from Wiley-Blackwell).

are shown in Table 5.7. Part of the argument in favour of factor 3 as a hydrogen-bonding factor is the negative loading of a proton donor stationary phase (HCM 18) and the positive loading of proton acceptor phases (DEGA, PPE5R, and EGA). Another part of the argument is that nonpolar stationary phases have approximately zero loadings with this factor.

The attempted physicochemical interpretation of the factors highlights a common problem with PCA and FA, along with the question of how significant is a factor (or PC) which only describes a few % of the data variance. The significance of factor 3 is uncertain but it is clearly useful since a scores plot of factor 3 versus factor 2 separates the solutes in terms of chemical functionality as shown in Figure 5.4.

The physiochemical interpretation of factors is nicely illustrated by a factor analysis of solvent parameters reported by Svoboda and co-workers [10]. Many attempts have been made to characterize solvents in terms of their effect on chemical reactions, their ability to dissolve solutes, their effect on properties such as spectra, and so on. This has led to the development of many different parameters and a variety of attempts have

Table 5.7   Factor loadings calculated by PFA (principal
factor analysis) method (reproduced from ref. [9] with
permission of The Pharmaceutical Society of Japan).

	Factor 1	Factor 2	Factor 3
CCR*	0.922	0.990	0.997
AP1-L	0.898	0.431	−0.001
CASTOR	0.983	0.133	−0.094
CW1000	0.947	−0.315	−0.004
DEGA	0.945	−0.311	0.083
D2EHS	0.968	0.238	−0.044
DIDP	0.969	0.224	−0.040
DC550	0.942	0.321	0.087
EGA	0.953	−0.279	0.089
HCM18	0.971	0.050	−0.227
HYP	0.901	−0.419	−0.052
IGE880	0.981	−0.183	−0.005
NPGA	0.994	−0.095	0.022
PPE5R	0.966	0.209	0.150
QUAD	0.931	−0.326	−0.089
SE30	0.931	0.359	0.000
SAIB	0.998	0.002	0.030
TCP	0.993	0.031	−0.016
TX305	0.982	−0.181	−0.055
U2000	0.994	−0.071	−0.037
VF50	0.912	0.399	0.020
XF1150	0.960	−0.195	0.138

*Cumulative contribution ratio.

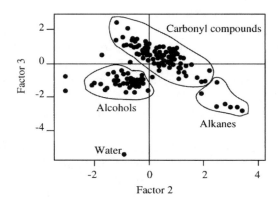

Figure 5.4   Scores plot on factors 3 and 2 derived from GC retention data (reproduced from ref. [9] with permission of The Pharmaceutical Society of Japan).

Table 5.8 Coordinates of parameters in factor space of the selected set (set 2) (reproduced from ref. [10] with permission of the Collection of Czechoslovak Communications).

Parameter	F_1	F_2	F_3	F_4
B	−0.079	0.012	0.610	0.008
$E_T(30)$	0.181	0.263	0.002	0.092
Z	0.305	−0.018	0.021	0.133
S_1[a]	0.261	0.078	0.144	0.014
S_2[b]	0.225	0.208	−0.010	0.051
DN	0.187	−0.208	0.560	−0.161
ε	0.154	0.237	0.026	−0.047
n_D^{20}	−0.010	0.044	−0.010	−0.600
YP[cc]	−0.097	0.473	0.090	0.068
pp[d]	−0.014	0.046	−0.008	−0.604
E	0.297	0.073	−0.042	0.124
a^{14N}	0.355	−0.015	0.022	0.037
AN	0.390	−0.066	−0.026	−0.014
π^*	0.026	0.413	−0.032	−0.250
$\log P$	−0.113	−0.103	−0.152	−0.167
δ	0.373	0.024	−0.014	−0.114
χ_R	0.094	−0.480	−0.046	0.106
δ^2	0.363	−0.011	−0.071	−0.113
β	−0.137	0.169	0.472	0.140
$^n\chi$[e]	0.028	−0.320	0.159	−0.222

[a]S_1, the parameter S defined by Zelinski.
[b]S_2, the parameter S defined by Brownstein.
[c]The Kirkwood function of dielectric function $YP = (\varepsilon - 1)/(2\varepsilon + 1)$.
[d]The function of refractive index $PP = (n^2 - 1)/(n^2 + 1)$.
[e]The index of molecular connectivity of the nth order.

been made to relate these parameters to one another. Table 5.8 shows the loadings of 20 parameters, for 51 solvents, on four factors which have been rotated by the varimax method. The parameters associated with the first factor describe electrophilic solvation ability, while those associated with factor 2 concern solvent polarity. The third factor is associated with nucleophilic solvation ability and the fourth factor with dispersion solvation forces.

It was proposed that a property, A, which is dependent on solvent effects could be described by an equation consisting of these four factors as shown in Equation (5.4):

$$A = A_o + aAP + bBP + eEP + pPP \tag{5.4}$$

where AP (acidity parameter) is the electrophilic factor, BP (basicity parameter) is the nucleophilic parameter, EP (electrostatic parameter)

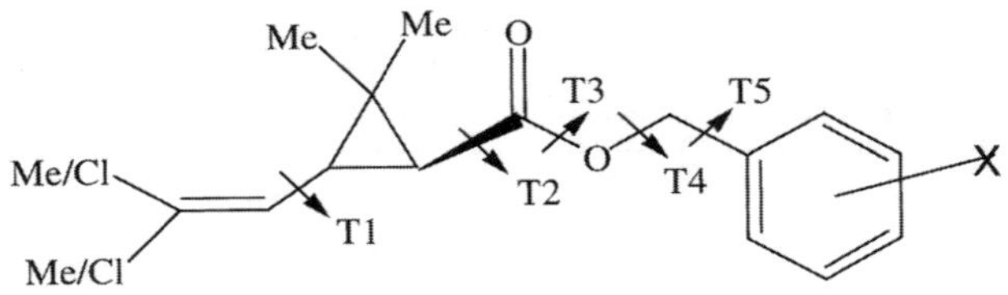

Figure 5.5 Pyrethroid parent structure; rotatable bonds are indicated by arrows (reproduced from ref. [12] with kind permission of Springer Science + Business Media).

the polar factor, and *PP* (polarizability parameter) the dispersion factor. A_0 is the value of the solvent-dependent property in a medium in which the solvent factors are zero (cyclohexane was suggested as a suitable solvent for this). The coefficients *a*, *b*, *e*, and *p* are fitted by regression (see Chapter 6) to a particular data set and represent the sensitivity of a process, as measured by the values of *A*, to the four solvent factors. Application of this procedure to 22 chemical data sets identified examples of processes with quite different dependencies on these solvent properties.

The final example of FA to be discussed here involves a number of insecticides, which are derivatives of the pyrethroid skeleton shown in Figure 5.5.

Computational chemistry methods were used to calculate a set of 70 molecular properties to describe these compounds [11]. Factor analysis identified a set of eight factors (Table 5.9) which explained 99 % of the variance in the chemical descriptor set. The physicochemical significance of the factors can be judged to some extent by an examination of the properties most highly associated with each factor, as shown in the table. The factors were shown to be of importance in the description of several biological properties of these compounds (see Section 8.5). These pyrethroid analogues are flexible as indicated by the rotatable bonds marked in Figure 5.5.

The physicochemical properties used to derive the factors shown in Table 5.9 were based on calculations carried out on a single conformation of each compound using a template from an X-ray crystal structure. In an attempt to take account of conformational flexibility, molecular dynamics simulations were run on the compounds and a number of representative conformations were obtained for each analogue [12]. The majority of these conformations represent an extended form of the molecule, similar to the X-ray template, but some are 'folded'. The physicochemical property calculations were repeated for each of the representative

Table 5.9 The molecular features of the QSAR pyrethroids (identified by FA) (reproduced from ref. [11] with permission of Wiley-Blackwell).

Factor	Principal associated descriptors and loadings	Molecular feature
1	A11(0.97), A12(0.99), A13(0.91), A16(−0.99) A17(−0.95), MW(−0.63)	The nature of the acid moiety indicated by associated MW and partial atomic charges
2	NS9(0.97), NS10(0.97) NS8(0.96), NS11(0.96) ES15(0.77), NS7(0.77)	Tendency of the atoms around the central ester linkage to accept and the *cis* geminal methyl to donate electrons
3	ES1(0.94), ES7(0.94) ES8(0.94), ES9(0.94) ES10(0.95)	Tendency of the atoms associated with the ester linkage to donate electrons
4	A3(0.84), A5(0.84) A10(0.85), ET(−0.84)	Partial atomic charges on the *meta* carbon atoms of the benzyl ring and the carbonyl carbon
5	NS2(0.90), NS3(0.75) NS5(0.77), NS6(0.87) A7(−0.77)	Tendency of the *ortho-* and *meta*-carbon atoms of the benzyl ring to accept electrons
6	DCA(0.86), SA(0.79) CD(0.71), VWV(0.71)	Molecular bulk, surface area and distance of closest approach
7	DVZ(0.82), DM(0.81)	Dipole strength and orientation
8	MW(0.70)	Molecular weight due to the alcohol moiety

conformations of each analogue and the resulting descriptors were averaged. Running factor analysis on this time-averaged set resulted in the identification of nine 'significant' factors (eigenvalues greater than 1), one more than the factor analysis of the static set. This additional factor suggests that there is extra information in the time-averaged set. Several of the static and time-averaged factors were highly correlated with one another and it was shown that these factors could be used to explain the lifetimes of the folded conformations.

Before leaving this description of factor analysis it is worth returning to another similarity with principal components analysis (and some other 'latent' variable methods, see Section 7.3.2) and that is the decision about the number of important or 'significant' or perhaps just useful factors/PC's to consider. One approach which is commonly employed is to construct what is known as a scree plot. A scree plot is simply a plot of the eigenvalues of the factors against their factor number. Since factors (and PC's) are extracted in order of their explanation of variance the eigenvalues decrease as the factor number increases and thus the

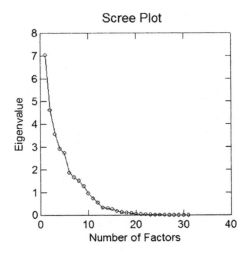

Figure 5.6 Scree plot from a factor analysis of 31 variables describing 35 compounds.

plot can look a little like the side of a mountain with irregularities as the eigenvalues change, hence the description scree plot. Such a plot is shown in Figure 5.6.

This scree plot shows three distinct discontinuities, or 'elbows', which indicate changes in the amount of variance described. The first of these happens at factor 5 where the next eigenvalue is considerably lower. The eigenvalues had been falling in quite a regular fashion for the first four factors, with the change in eigenvalue between four and five smaller, so this might be a natural stopping point. The next discontinuity occurs at factor 8 and as this and the next factor both have eigenvalues greater than 1 this might also be a natural stopping point. As mentioned in the previous example, an eigenvalue of 1 may be used as a cutoff value since with an autoscaled data set each variable has a variance of 1, therefore a factor or principal component with an eigenvalue less than 1 is explaining less variance than one of the original variables. This is a commonly used 'stopping rule' for factor analysis and PCA. The last discontinuity occurs at factor 13 with the eigenvalues for the subsequent factors falling in a regular fashion until they vanish to almost nothing. So, the scree plot suggest three possible stopping points and the eigenvalue > 1 rule suggests 1 stopping point which coincides with a point identified with the scree plot. Unfortunately, neither of these approaches ensures a 'correct' choice as demonstrated in Section 7.3.1.

5.4   CLUSTER ANALYSIS

Chatfield and Collins [5], in the introduction to their chapter on cluster analysis, quote the first sentence of a review article on cluster analysis by Cormack [13]: 'The availability of computer packages of classification techniques has led to the waste of more valuable scientific time than any other "statistical" innovation (with the possible exception of multiple-regression techniques).' This is perhaps a little hard on cluster analysis and, for that matter, multiple regression but it serves as a note of warning. The aim of this book is to explain the basic principles of the more popular and useful multivariate methods so that readers will be able to understand the results obtained from the techniques and, if interested, apply the methods to their own data. This is *not* a substitute for a formal training in statistics; the best way to avoid wasting one's own valuable scientific time is to seek professional help at an early stage.

Cluster analysis (CA) has already been briefly mentioned in Section 2.3, and a dendrogram was used to show associations between variables in Section 3.6. The basis of CA is the calculation of distances between objects in a multidimensional space using an equation such as Equation (5.1). These distances are then used to produce a diagram, known as a dendrogram, which allows the easy identification of groups (clusters) of similar objects. Figure 5.7 gives an example of the process for a very simple two-dimensional data set.

The two most similar (closest) objects in the two-dimensional plot in part (a) of the figure are A and B. These are joined together in the dendrogram shown in part (b) of the figure where they have a low value of dissimilarity (distance between points) as shown on the scale. The similarity scale is calculated from the interpoint distance matrix by finding the minimum and maximum distances, setting these equal to some arbitrary scale numbers (e.g. 0 and 1), and scaling the other distances to lie between these limits. The next smallest interpoint distance is between point C and either A or B, so this point is joined to the A/B cluster. The next smallest distance is between D and E so these two points form a cluster and, finally, the two clusters are joined together in the dendrogram. This process is hierarchical and the links between clusters have been single; the procedure is known, unsurprisingly, as single-link hierarchical cluster analysis and is one of the most commonly used methods. Another point to note from this description of CA is that clusters were built up from individual points, the process is agglomerative. CA can

(a)

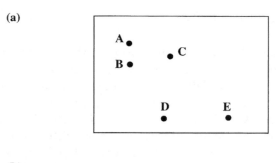

(b)

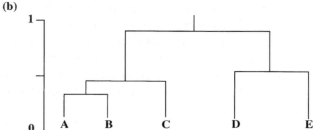

Figure 5.7   Illustration of the production of a dendrogram for a simple two-dimensional data set (reproduced with permission from ref. [3] with permission of Energia Nuclear & Agricultura).

start off in the other direction by taking a single cluster of all the points and splitting off individual points or clusters, a divisive process.

There are many different ways in which clusters can be generated; all of the examples that will be described in this section use the agglomerative, hierarchical, single-linkage method, usually referred to as 'cluster analysis'. Most textbooks of multivariate analysis have a chapter describing some of the alternative methods for performing CA, and Willett [14] deals with chemical applications. It may have been noticed that in this description of CA the points to be clustered were referred to as just that, points in a multidimensional space. They have not been identified as samples or variables since CA, like many multivariate methods, can be used to examine relationships between samples or variables. For the former we can view the data set as a collection of n objects in a p-dimensional parameter space. For the latter we can imagine a data set 'turned on its side' so that it is a collection of p objects in an n-dimensional sample space. When using CA to examine the relationships between variables, the distance measure employed is often the correlation coefficients between variables, as shown in Figure 3.5.

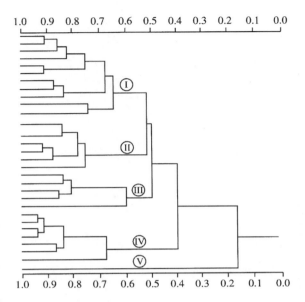

Figure 5.8 Dendrogram of water samples characterized by their concentrations of Ca, K, Na, and Si (reproduced from ref. [15] with permission of Wiley-Blackwell).

The study of mineral waters characterized by elemental analysis discussed in Section 5.2 [3] provides a nice example of the use of CA to classify samples. Figure 5.8 shows a dendrogram of water samples from one geographical region (Lindoya) described by the concentrations of four elements. The water samples were drawn from six different locations in this region and one group on the dendrogram, cluster IV, contained all the samples from one of these locations. The samples from the other five locations are contained in clusters I, II, and III. One sample, cluster V, is clearly an outlier from this set and thus must be subject to suspicion.

The characterization of fruit juices by various analytical measurements was used as an example of a principal component scores plot (Figure 4.9) in Chapter 4 [15]. A dendrogram from this data is shown in Figure 5.9 where it is clearly seen that the grape, apple, and pineapple juice samples form distinct clusters. The apple and pineapple juice clusters are grouped together as a single cluster which is quite distinct from the cluster of grape juice samples. This is interesting in that it mimics the results of the PCA; on the scores plot, all three groups are separated, but the first component mainly serves to separate the grape juices from the others while the second component separates apple and pineapple juices. This is a good illustration of the way that different multivariate

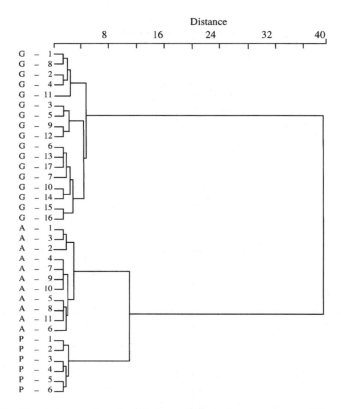

Figure 5.9 Dendrogram showing the associations between grape (G), apple (A), and pineapple (P) juice samples described by 15 variables (reproduced from ref. [16] with permission of Arzneimettel-Forschung).

methods tend to produce complementary and consistent views of the same data set.

The dendrogram in Figure 5.10 is derived from a data matrix of ED_{50} values for 40 neuroleptic compounds tested in 12 different assays in rats [16]. This is an example of a situation in which the data involves multiple dependent variables (see Chapter 8), but here the multiple biological data is used to characterize the tested compounds. The figure demonstrates that the compounds can be split up into five clusters with three compounds falling outside the clusters. Compounds within a cluster would be expected to show a similar pharmacological profile and, of course, there is the finer detail of clusters within the larger clusters. A procedure such as this can be very useful when examining new potential drugs. If the pharmacological profile of a new compound can be matched to that

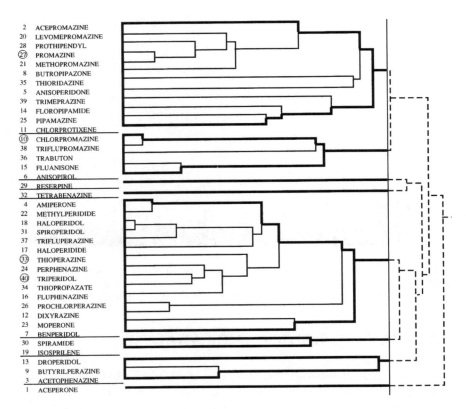

Figure 5.10   Dendrogram of the relationships between neuroleptic drugs characterized by 12 different biological tests (reproduced from ref. [12] with kind permission of Springer Science + Business Media).

of a marketed compound, then the early clinical investigators may be forewarned as to the properties they might expect to see.

The final example of a dendrogram to be shown here, Figure 5.11, is also one of the largest. This figure shows one thousand conformations of an insecticidal pyrethroid analogue (see Figure 5.5) described by the values of four torsion angles [12]. A dendrogram such as this was used for the selection of representative conformations from the one thousand conformations produced by molecular dynamics simulation. Conformations were chosen at equally spaced intervals across the dendrogram ensuring an even sampling of the conformational space described by the torsion angles. In fact, the procedure is not as simple as this and various approaches were employed (see reference for details) but sampling at even intervals was shown to be suitable.

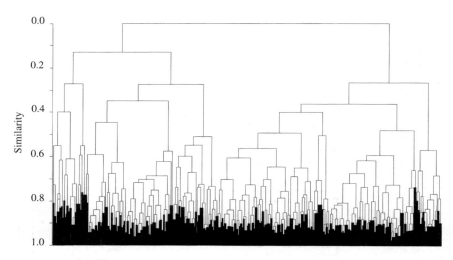

Figure 5.11 Dendrogram of the relationships of different conformations of a pyrethroid derivative described by the values of four torsion angles (reproduced from ref. [17] with permission from the American Chemical Society).

5.5   CLUSTER SIGNIFICANCE ANALYSIS

The advantage of unsupervised learning methods is that any patterns that emerge from the data are dependent on the data employed. There is no intervention by the analyst, other than to choose the data in the first place, and there is no attempt by the algorithm employed to 'fit' a pattern to the data, or seek a correlation, or produce a discriminating function (see Chapter 7). Any groupings of points which are seen on a non-linear map, a principal components plot, a dendrogram, or even a simple bivariate plot are solely due to the disposition of samples in the parameter space and it is unlikely, although not impossible, to have happened by chance. There is, however, a major drawback to the unsupervised learning approach and that is an evaluation of the quality or 'significance' of any clusters of points. Many analytical methods, particularly the parametric techniques based on assumptions about population distributions, have significance tests built in. If we look at the principal component scores plot for the fruit juices (Figure 4.9) or the dendrogram for the same data (Figure 5.8) it seems obvious that the groupings have some 'significance', but is this always the case? Is it possible to judge the quality of some unsupervised picture? McFarland and Gans [17] addressed this problem by means of a method which they termed *cluster significance analysis* (CSA). The concept underlying this method is quite

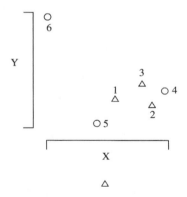

Figure 5.12 Plot of active (Δ) and inactive (○) compounds described by two parameters (reproduced from ref. [17] with permission of the American Chemical Society).

simple: for a given display of N samples which contains a cluster of M active (or otherwise interesting) samples, how 'tight' is the cluster of M samples compared with all the other possible clusters of M samples? Various measures of tightness could be used but the one chosen was the mean squared distance (MSD) which involves taking the sum of the squared distances between each pair of points in the cluster divided by the number of points in the cluster (M).

The process is nicely illustrated by a hypothetical example from the original report. Figure 5.12 shows a two-dimensional plot of six compounds, three active and three inactive. The total squared distance (TSD) for the active cluster is given by

$$TSD = (x_1 - x_2)^2 + (y_1 - y_2)^2 + (x_1 - x_3)^2 + (y_1 - y_3)^2 \\ + (x_2 - x_3)^2 + (y_2 - y_3)^2 \tag{5.4}$$

and the mean squared distance

$$MSD = {}^{TSD}\!/_3. \tag{5.5}$$

The probability that a cluster as tight as the active cluster would have arisen by chance involves the calculation of MSD for all the other possible clusters of three compounds. The number of clusters with an MSD value equal to or less than the active MSD is denoted by A (including the active cluster) and a probability is calculated as

$$p = {}^{A}\!/_N \tag{5.6}$$

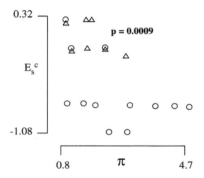

Figure 5.13 Plot of active (Δ) and inactive ($\circ$) inhibitors of monoamine oxidase (from ref. [17] copyright (1986) American Chemical Society).

where N is the total number of possible clusters of that size, in this case three compounds. It is obvious from inspection of the figure that there is one other cluster as tight as or tighter than the active cluster (compounds 2, 3, and 4) and that all other clusters have larger MSD values since they include compounds 1, 5, or 6. There are 20 possible clusters of three compounds in this set and thus $A = 2$, $N = 20$, and

$$p = {}^2\!/_{20} = 0.10 \tag{5.7}$$

If a probability level of 0.05 or less (95 % certainty or better) is taken as a significance level then this cluster of actives would be regarded as fortuitous.

Figure 5.13 shows a plot of a set of inhibitors of the enzyme mono-amine oxidase (MAO) described by steric (E_s^c) and hydrophobic (π) parameters. It can be seen that the seven active compounds mostly cluster in the top left-hand quadrant of the plot. The original data set involved a dummy parameter, D, to indicate substitution by OCH_3 or OH at a particular position, and in the application of CSA to this problem, a set of random numbers, RN, was added to the data. The results of CSA analysis for this data are shown in Table 5.10 where it is seen that lowest probability of fortuitous clustering is given by the combination of π and E_s^c.

This illustrates another feature of CSA; not only can it be used to judge the significance of a particular set of clusters, it can also be used to test the effect (on the tightness of clusters) of adding or removing a particular descriptor. Thus, it may be used as a selection criterion for the usefulness of parameters. One thing that should be noted from the table is the large number of possible subsets (77 520) that can be generated

Table 5.10 Application of CSA to a set of 20 MAO inhibitors (reproduced from ref. [17] with permission of the American Chemical Society).

Parameters	A*	p
D	21464	0.27688
RN	14825	0.19124
πc	1956	0.02523
E_s	118	0.00152
D, π	1299	0.01676
D, E_s^c	1175	0.01516
RN, E_s^c	172	0.00222
π, E_s^c	71	0.00092
RN, π, E_s^c	151	0.00195
D, π, E_s^c	78	0.00101

*From a total possible set of 77,520 subsets of 7.

for this data set. This may cause problems in the analysis of larger data sets in terms of the amount of computer time required. An approach to solving this problem is to compute a random sample of the possible combinations rather than exhaustively examining them all [17]. CSA has been compared with three other QSAR techniques in the analysis of three different data sets [18].

5.6 SUMMARY

Unsupervised learning methods, like the display techniques described in Chapter 4, are very useful in the preliminary stages of data analysis. Cluster analysis and FA produce easily understood displays from high-dimensional data sets and may be used when the number of variables in the set exceeds the number of samples. Although care must be exercised in the choice of class members when using k-nearest-neighbours, this and other methods described in this chapter should be reasonably safe from the danger of chance correlations. Cluster significance analysis allows us to attempt to assign significance levels to any 'interesting' groupings of samples seen using these methods or multivariate display techniques. Finally, in common with all of the other methods described in this book, it is not possible to say that any one technique is 'best'.

In this chapter the following points were covered:

1. classification by k-nearest neighbours;
2. factor analysis – similarity and differences with PCA;

3. use of factor analysis to see the relationships between variables;
4. use of factor analysis to visualize samples;
5. use of scree plots to choose 'significant' factors;
6. cluster analysis to examine relationships between samples;
7. cluster significance analysis to judge the 'quality' of clusters.

REFERENCES

[1] Kowalski, B.R. and Bender, C.F. (1972). *Analytical Chemistry*, **44**, 1405–11.
[2] Chu, K.C., Feldman, R.J., Shapur, M.B., Hazard, G.F., and Geran, R.I. (1975). *Journal of Medicinal Chemistry*, **18**, 539–45.
[3] Scarminio, I.S., Bruns, R.E., and Zagatto, E.A.G. (1982). *Energia Nuclear e Agricultura*, **4**, 99–111.
[4] Goux, W.J. and Weber, D.S. (1993). *Carbohydrate Research*, **240**, 57–69.
[5] Chatfield, C. and Collins, A.J. (1980). *Introduction to Multivariate Analysis*. Chapman & Hall, London.
[6] Malinowski, E.R. (1991). *Factor Analysis in Chemistry*. John Wiley & Sons, Inc., New York.
[7] Jackson, J.E. (1991). *A User's Guide to Principal Components*. John Wiley & Sons, Inc., New York.
[8] Li-Chan, E., Nakai, S., and Wood, D.E. (1987). *Journal of Food Science*, **52**, 31–41.
[9] Takagi, T., Shindo, Y., Fujiwara, H., and Sasaki, Y. (1989). *Chemical and Pharmaceutical Bulletin*, **37**, 1556–60.
[10] Svoboda, P., Pytela, O., and Vecera, M. (1983). *Collection of Czechoslovak Chemical Communications*, **48**, 3287–306.
[11] Ford, M.G., Greenwood, R., Turner, C.H., Hudson, B., and Livingstone, D.J. (1989). *Pesticide Science*, **27**, 305–26.
[12] Hudson, B.D., George, A.R., Ford, M.G., and Livingstone, D.J. (1992). *Journal of Computer-aided Molecular Design*, **6**, 191–201.
[13] Cormack, R.M. (1971). *Journal of the Royal Statistical Society*, **A134**, 321–67.
[14] Willett, P. (1987). *Similarity and Clustering in Chemical Information Systems*. Research Studies Press, John Wiley & Sons, Ltd, Chichester.
[15] Dizy, M., Martin-Alvarez, P.J., Cabezvdo, M.D., and Polo, M.C. (1992). *Journal of the Science of Food and Agriculture*, **60**, 47–53.
[16] Lewi, P.J. (1976). *Arzneimettel-Forschung*, **26**, 1295–1300.
[17] McFarland, J.W. and Gans, D.J. (1986). *Journal of Medicinal Chemistry*, **29**, 505–14.
[18] McFarland, J.W. and Gans, D.J. (1987). *Journal of Medicinal Chemistry*, **30**, 46–9.

6

Regression Analysis

Points covered in this chapter

- Simple linear regression
- Multiple linear regression
- Constructing multiple regression models
- Model validation and chance effects
- Non-linear regression
- Regression with indicator variables (Free and Wilson)

6.1 INTRODUCTION

Regression analysis is one of the most commonly used analytical methods in chemistry, including all of its specialist subdivisions and allied sciences. Indeed, the same can probably be said about most forms of science. The reason for its appeal lies perhaps in the fact that the method formalizes something that the human pattern recognizer does instinctively, and that is to fit a line or a curve through a set of data points. We are accustomed to looking for trends in the data that the world presents to us, whether it be unemployment or inflation figures, or the results of some painstakingly performed experiments. We do this in the hope, or expectation, that the trends will reveal some underlying explanation of how or why the data is produced. In its simplest form, regression analysis involves fitting a straight line through a set of data points represented by just two variables, calculating an equation for the fitted line, and providing estimates of how well the points fit the line. The first section of this chapter will

A Practical Guide to Scientific Data Analysis   David Livingstone
© 2009 John Wiley & Sons, Ltd

discuss simple linear regression and the calculation and interpretation of its statistics. The next section describes multiple linear regression: how the equations are constructed, non-linear regression models and the use of indicator variables in regression, including Free and Wilson analysis. The final section discusses some important features of regression analysis such as the comparison of regression models, tests for robustness, and the problems of chance correlations. Regression analysis based on variables derived from multivariate data, principal components, factors, and latent variables is discussed in Chapter 7, Supervised Learning.

6.2   SIMPLE LINEAR REGRESSION

We have already seen in Chapter 1 an example of a simple linear regression model (Equation (1.6), Figure 1.8) in which anaesthetic activity was related to the hydrophobicity parameter, π. How was the equation derived? If we consider the data shown plotted in Figure 6.1, it is fairly obvious that a straight line can be fitted through the points.

A line is shown on the figure and is described by the well-known equation for a straight line.

$$y = mx + c \qquad (6.1)$$

The value of c (2.0), the intercept of the line, can be read from the graph where $x = 0$ ($y = m0 + c$) and the value of m (1.0), the slope of the line, by taking the ratio of the differences in the y and x values at two points on the line $(y_2 - y_1)/(x_2 - x_1)$. A line such as this can be obtained

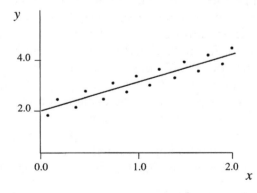

Figure 6.1   Plot of the values of variable y against variable x with a fitted straight line.

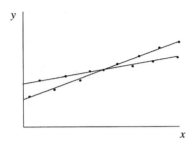

Figure 6.2   Plot of y against x where two different straight lines can be fitted to the points.

easily by laying a straight edge along the data points and it is clear that for this data, if another person repeated the procedure, a line with a very similar equation would result.

Figure 6.2 shows a different situation in which the data points still clearly correspond to a straight line but here it is possible to draw different lines through the data. Which of these two lines is best? Is it possible to say that one line is a better fit to the data than the other, or is some other line the best fit? Whether or not there is some way of saying what the 'right' answer is, it is clear that some objective way of fitting a line to data such as those shown in these figures is required. One such technique is called the method of least squares, or ordinary least squares (OLS), in which the squares of the distances between the points and the line are minimized.

This is shown in Figure 6.3 for the same data points as Figure 6.2 with the exception that there is an extra data point in this figure. The extra point corresponds to the mean of the x ($\overline{x}$) and y ($\overline{y}$) data,

$$\overline{x} = \frac{\sum\limits_{i=1}^{n} x_i}{n}, \quad \overline{y} = \frac{\sum\limits_{i=1}^{n} y_i}{n} \tag{6.2}$$

and it can be seen that the regression line, or least-squares line, passes through this point. Since the point $(\overline{x}, \overline{y})$ lies on the line, the equation can be written as

$$y - \overline{y} = m(x - \overline{x}) \tag{6.3}$$

the constant term, c, having disappeared since it is explained by the means

$$c = \overline{y} - m\overline{x} \tag{6.4}$$

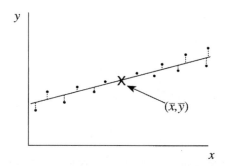

Figure 6.3   Illustration of the process of least squares fitting.

Equation (6.3) can be rewritten as

$$y = \overline{y} + m(x - \overline{x}) \tag{6.5}$$

and thus for any value of x (x_i) an estimate of the y value ($\hat{y}_i$) can be made

$$\hat{y}_i = \overline{y} + m(x_i - \overline{x}) \tag{6.6}$$

The error in prediction for the y value corresponding to this x value is given by

$$y_i - \hat{y}_i = y_i - \overline{y} - m(x_i - \overline{x}) \tag{6.7}$$

This equation can be used to express a set of errors for the prediction of y values over the whole set of data (n pairs of points) and the sum of the squares of these errors is given by

$$U = \sum_{i=1}^{n} (y_i - \overline{y} - m(x_i - \overline{x}))^2 \tag{6.8}$$

Minimization of this sum of squares gives the slope of the regression line (m), which is equivalent to minimizing the lengths of the dotted lines, shown in Figure 6.3, between the data points and the fitted line. It can be shown for minimum U (where $dU/dm = 0$ and d^2U/dm^2 is positive) that the slope is given by

$$m = \frac{\sum\limits_{i=1}^{n}(x_i - \overline{x})(y_i - \overline{y})}{\sum\limits_{i=1}^{n}(x_i - \overline{x})^2} \tag{6.9}$$

Table 6.1 Assumptions for simple linear regression.

1	The x and y data is linearly related
2	The error is in y, the dependent variable
3	The average of the errors is 0
4	The errors are independent; there is no serial correlation among the errors (knowing the error for one observation gives no information about the others)
5	The errors are of approximately the same magnitude
6	The errors are approximately normally distributed (around a mean of zero)

Thus both the slope and the intercept of the least-squares line can be calculated from simple sums using Equations (6.4) and (6.9). In practice, few people ever calculate regression lines in this way as even quite simple scientific calculators have a least-squares fit built in. However, it is hoped that this brief section has illuminated the principles of the least-squares process and has shown some of what goes on in the 'black box' of regression packages.

Having fitted a least-squares line to a set of data points, the question may be asked, 'How well does the line fit?' Before going on to consider this, it is necessary to state some of the assumptions, hitherto unmentioned, that are implicit in the process of regression analysis and which should be satisfied for the linear regression model to be valid. These assumptions are summarized in Table 6.1, and, in principle, all of these assumptions should be tested before regression analysis is applied to the data.

In practice, of course, few if any of these assumptions are ever checked but if simple linear regression is found to fail when applied to a particular data set, it may well be that one or more of these assumptions have been violated. Assumptions 1 and 2 are particularly important since the data should look as though it is linearly related and at least the majority of the error should be contained in the y variable (called a regression of y on x). In many chemical applications this latter assumption will be quite safe as the dependent variable will often be some experimental quantity whereas the descriptor variables (the x set) will be calculated or measured with good precision.

The assumption of a normal distribution of the errors allows us to put confidence limits on the fit of the line to the data. This is carried out by the construction of an analysis of variance table (the basis of many statistical tests) in which a number of sums of squares are collected.[1]

---

[1] There is no agreed convention for abbreviating these sums of squares, other treatments may well use different sets of initials.

The total sum of squares (TSS), in other words the total variation in y, is given by summation of the difference between the observed y values and their mean.

$$TSS = \sum_{i=1}^{n} (y_i - \bar{y})^2 \qquad (6.10)$$

This sum of squares is made up from two components: the variance in y that is explained by the regression equation (known as the explained sum of squares, ESS), and the residual or unexplained sum of squares, RSS. The ESS is given by a comparison of the predicted y values ($\hat{y}$) with the mean

$$ESS = \sum_{i=1}^{n} (\hat{y}_i - \bar{y})^2 \qquad (6.11)$$

and the RSS by comparison of the actual y values with the predicted

$$RSS = \sum_{i=1}^{n} (y_i - \hat{y}_i)^2 \qquad (6.12)$$

The total sum of squares is equal to these two sums

$$TSS = ESS + RSS \qquad (6.13)$$

These sums of squares are shown in the analysis of variance (ANOVA) table (Table 6.2). The mean squares are obtained by division of the sums of squares by the appropriate degrees of freedom. One degree of freedom is 'lost' with each parameter calculated from a set of data so the total sum of squares has $n - 1$ degrees of freedom (where n is the number of data points) due to calculation of the mean. The residual sum of squares has

Table 6.2   ANOVA table.

Source of variation	Sum of squares	Degrees of freedom	Mean square
Explained by regression	ESS	1	MSE (=ESS)
Residual	RSS	$n - 2$	MSR (=RSS/$n - 2$)
Total	TSS	$n - 1$	MST (=TSS/$n - 1$)

$n - 2$ degrees of freedom due to calculation of the mean and the slope of the line. The explained sum of squares has one degree of freedom corresponding to the slope of the regression line.

Knowledge of the mean squares and degrees of freedom allows assessment of the significance of a regression equation as described in the next section, but how can we assess how well the line fits the data? Perhaps the best known and most misused regression statistic is the correlation coefficient. The squared multiple correlation coefficient (r^2) is given by division of the explained sum of squares by the total sum of squares

$$r^2 = \frac{\text{ESS}}{\text{TSS}}. \tag{6.14}$$

This can take a value of 0, where the regression is explaining none of the variance in the data, up to a value of 1 where the regression explains all of the variance in the set. r^2 multiplied by 100 gives the percentage of variance in the data set explained by the regression equation. The squared correlation coefficient is the square of the simple correlation coefficient, r, between y and x (see Box 2.1 in Chapter 2, p. 39). This correlation coefficient can take values between -1, a perfect negative correlation (y decreases as x increases), and $+1$, a perfect positive correlation. Correlation coefficients, both simple and multiple (where several variables are involved), can be very misleading. Consider the data shown in Figure 6.4.

Part (a) of the figure shows a set of data in which y is clearly dependent on x by a simple linear relationship; part (b) shows two separate 'clouds' of points where the line has been fitted between the two groups; parts (c) and (d) show two situations in which a single rogue point has greatly affected the fit of the line.[2] Table 6.3 gives the data used to produce these plots and some of the statistics for the fit of the line. The correlation coefficients for these four graphs (and parts (e) and (f) of Figure 6.4) are very similar, as are the regression coefficients for x (0.55 to 0.82). There is a somewhat wider range in values for the constant term (0.20 to 0.80), but overall the statistics give little indication of the four different situations shown in parts (a) to (d) of the figure. Parts (e) and (f) show two other types of data set for which a single straight line fit is inappropriate,

---

[2] This situation is often referred to as a 'point and cluster effect'; the regression line is fitted effectively between two points, the rogue point and the cluster of points making up the rest of the set.

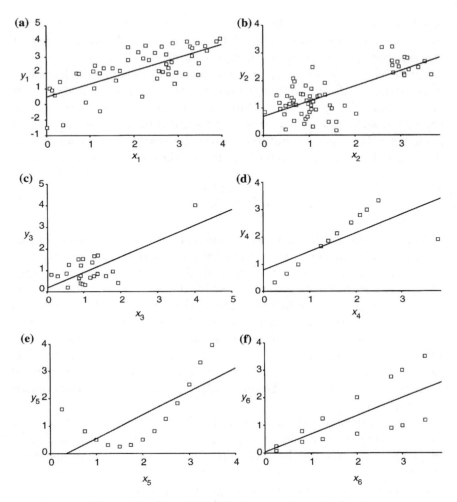

Figure 6.4 Plot of six different sets of y and x data.

a curve (e) and two separate straight lines (f). Once again the statistics give little warning, although the large standard errors of the constants do suggest that something is wrong. The lesson from this is clear; it is not possible to assess goodness of fit simply from a correlation coefficient. Indeed, this statistic (and others) can be very misleading, as well as very useful. In the case of a simple problem such as this involving just two variables the construction of a bivariate plot will reveal the patterns in the data. More complex situations present extra problems as discussed in the next section and Section 6.4.

Table 6.3 Data and statistics for the regression of y on x for the six data sets in Figure 6.4 (after an original example by Anscombe [1]).[*]

y_1	x_1	y_1	x_1	y_2	x_2	y_2	x_2
1.898	2.790	2.092	2.573	0.920	1.149	1.090	0.716
3.318	2.111	2.820	1.879	1.138	0.577	1.187	0.648
3.385	3.672	2.287	1.318	1.266	1.047	1.902	1.343
−0.460	1.224	2.011	2.967	1.267	0.942	1.960	0.683
0.900	0.130	2.819	2.354	0.513	0.673	1.088	0.676
3.718	2 276	2.127	1.686	0.401	0.892	0.765	0.462
2.046	2 668	1.866	3.462	0.770	0.831	2.081	0.654
−1.344	0.374	3.655	2.644	0.483	1.422	1.783	0.637
2.459	1.114	3.630	1.880	1.258	0.661	1.178	0.350
3.559	3.333	1.020	0.070	0.594	0.922	1.065	1.779
3.667	3.631	3.865	2.897	0.671	0.962	1.069	0.485
2.909	2.224	1.986	0.689	1.702	1.055	1.141	0.977
0.589	0.198	1.948	0.749	1.420	1.017	1.414	1.188
4.150	3.973	1.442	0.312	0.204	0.483	2.484	1.074
3.066	3.326	−1.505	0.007	1.453	0.295	0.837	0.036
3.859	3.459	2.590	3.483	1.355	0.617	0.835	1.004
2.093	1.082	2.664	2.866	1.432	0.521	2.170	3.090
3.647	3.294	1.989	1.230	1.810	0.927	2.500	2.970
3.255	2.487			0.776	2.026	3.220	2.830
1.014	1.101			1.500	0.819	2.670	2.940
1.639	2.532			0.934	1.055	2.800	3.160
0.484	2.195			1.219	1.090	2.230	3.080
1.294	2.928			1.076	1.043	2.680	3.540
1.521	1.595			0.962	1.489	2.710	2.820
2.554	2.750			1.449	1.227	2.200	3.680
3.170	2.751			0.957	0.372	2.830	3.100
3.962	3.893			1.486	0.865	2.540	2.830
2.304	2.793			1.084	1.452	2.380	3.190
1.898	3.172			0.483	1.589	2.470	3.400
0.122	0.903			1.477	1.343	3.200	2.600
0 309	1.523			0.168	1.593	2.280	2.840
3.979	3.218			0.307	1.080		
		r = 0.702				r = 0.706	
		F = 46.77				F = 60.55	
		SE = 0.95				SE = 0.56	
		RC = 0.82 (0.12)				RC = 0.55 (0.07)	
		c = 0.48 (0.12)				c = 0.69 (0.12)	

y_3	x_3	y_4	x_4	y_5	x_5	y_6	x_6
0.191	0.542	0.333	0.250	1.600	0.250	0.250	0.250
1.270	0.575	0.665	0.500	0.813	0.750	0.800	0.800
1.536	0.961	0.998	0.750	0.500	1.000	1.250	1.250
0.943	1.772	1.663	1.250	0.313	1.250	2.000	2.000
0.742	0.895	1.862	1.400	0.250	1.500	2.750	2.750
0.837	1.361	2.128	1.600	0.313	1.750	3.000	3.000
0.851	0.517	2.527	1.900	0.500	2.000	3.500	3.500
0.399	0.900	2.793	2.100	0.813	2.250	0.080	0.250

(continued)

Table 6.3 (Continued)

y_3	x_3	y_4	x_4	y_5	x_5	y_6	x_6
1.369	1.215	2.993	2.250	1.250	2.500	0.400	0.800
0.643	0.854	3.325	2.500	1.813	2.750	0.510	1.250
0.740	1.590	1.900	3.800	2.500	3.000	0.700	2.000
0.405	1.914			3.313	3.250	0.910	2.750
0.731	0.275			3.950	3.500	0.995	3.000
0.733	1.262					1.195	3.500
0.807	0.095						
1.512	0.847						
1.233	0.903						
0.666	1.164						
0.619	0.850						
0.365	0.948						
1.692	1.365						
0.319	1.015						
1.667	1.268						
4.000	4.000						
r = 0.702		r = 0.709		r = 0.707		r = 0.708	
F = 21.34		F = 9.11		F = 10.97		F = 12.09	
SE = 0.56		SE = 0.72		SE = 0.90		SE = 0.80	
RC = 0.72 (0.16)		RC = 0.68 (0.22)		RC = 0.85 (0.26)		RC = 0.66 (0.19)	
c = 0.20 (0.21)		c = 0.80 (0.43)		c = -0.30 (0.57)		c = 0.04 (0.42)	

[*]The statistics reported for each fit are – the simple correlation coefficient, r; the F statistic, F; the standard error of the fit, SE; the regression coefficient for X, RC, followed by its standard error in brackets; the constant of the equation, c, followed by its standard error in brackets.

6.3 MULTIPLE LINEAR REGRESSION

Multiple linear regression is an extension of simple linear regression by the inclusion of extra independent variables

$$y = ax_1 + bx_2 + cx_3 + \ldots + \text{constant} \tag{6.15}$$

Least squares may be used to estimate the regression coefficients (a, b, c, and so on) for the independent variables (x_1, x_2, x_3, and so on), and the value of the constant term. Goodness of fit of the equation to the data can be obtained by calculation of a multiple correlation coefficient (R^2) just as for simple linear regression. In the case of simple linear regression it is easy to see what the fitting procedure is doing, i.e. fitting a line to the data, but what does multiple regression fitting do? The answer is that multiple regression fits a surface. Figure 6.5 shows the surface fitted by a

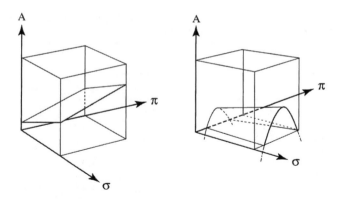

Figure 6.5 Illustration of different surfaces corresponding to two-term regression equations.

two-term equation in π and σ (a plane) and an equation which includes a squared term.

It is difficult to illustrate the results of fitting higher order equations but the principle is the same; multiple regression equations fit a surface to data with a dimensionality equal to the number of independent variables in the equation. It was shown in the previous section that the correlation coefficient can be a misleading statistic for simple linear regression fitting and the same is true for the multiple regression case. It is more difficult (or impossible) to check a multiple regression fit by plotting the data points with respect to all of the parameters in the equation, but one way that even the most complicated regression model can be evaluated is by plotting predicted y values against the observed values. If the regression equation is a perfect fit to the data ($R^2 = 1$), then a plot of the predicted versus observed should give a straight line with a slope of one and an intercept of zero. When some particular points are badly predicted it will be obvious from this plot; a curved plot suggests some other equation is more appropriate.

What about an assessment of the significance of the fit of a multiple regression equation (or simple regression) to a set of data? A guide to the overall significance of a regression model can be obtained by calculation of a quantity called the F statistic. This is simply the ratio of the explained mean square (MSE) to the residual mean square (MSR)

$$F = \frac{\text{MSE}}{\text{MSR}}. \tag{6.16}$$

Table 6.4 5 % points (95 % confidence) of the F-distribution (reproduced from ref. [2] copyright Cambridge University Press).

				v_1			
v_2	1	2	3	4	5	10	∞
1	161.4	199.5	215.7	224.6	230.2	241.9	254.3
2	18.5	19.0	19.2	19.2	19.3	19.4	19.5
3	10.13	9.55	9.28	9.12	9.01	8.79	8.53
4	7.71	6.94	6.59	6.39	6.26	5.96	5.63
5	6.61	5.79	5.41	5.19	5.05	4.74	4.36
6	5.99	5.14	4.76	4.53	4.39	4.06	3.67
7	5.59	4.74	4.35	4.12	3.97	3.64	3.23
8	5.32	4.46	4.07	3.84	3.69	3.35	2.93
9	5.12	4.26	3.86	3.63	3.48	3.14	2.71
10	4.96	4.10	3.71	3.48	3.33	2.98	2.54
15	4.54	3.68	3.29	3.06	2.90	2.54	2.07
20	4.35	3.49	3.10	2.87	2.71	2.35	1.84
30	4.17	3.32	2.92	2.69	2.53	2.16	1.62
40	4.08	3.23	2.84	2.61	2.45	2.08	1.51
∞	3.84	3.00	2.60	2.37	2.21	1.83	1.00

An F statistic is used by looking up a standard value for F from a table of F statistics and comparing the calculated value with the tabulated value. If the calculated value is greater than the tabulated value, the equation is significant at that particular confidence level. F tables normally have values listed for different levels of significance, e.g. 10 %, 5 %, and 1 %. As might be expected, the F values are greater for higher levels of significance. This is equivalent to saying that we expect the explained mean square to be even larger than the residual mean square in order to have a higher level of confidence in the fit. This seems like good common sense! Table 6.4 gives some values of the F statistic for different numbers of degrees of freedom at a significance level of 5 %. It can be seen that the table has entries for two degrees of freedom, the rows and the columns. These correspond to the number of degrees of freedom associated with the explained mean square, MSE, which is given by p (where p is the number of independent variables in the equation) and with the residual mean square, MSR, which is given by $n-p-1$ (where n is the number of data points). An F statistic is usually quoted as $F(v_1 v_2)$, where $v_1 = p$ and $v_2 = n-p-1$. When regression equations are reported, it is not unusual to find the appropriate tabulated F value quoted for comparison with the calculated value.

The squared multiple correlation coefficient gives a measure of how well a regression model fits the data and the F statistic gives a measure

Table 6.5 Percentage points of the t distribution (reproduced from ref. [2] copyright Cambridge University Press).

				P			
v_1	25	10	5	2	1	0.2	0.1
	2.41	6.31	12.71	31.82	63.66	318.3	636.6
2	1.60	2.92	4.30	6.96	9.92	22.33	31.60
3	1.42	2.35	3.18	4.54	5.84	10.21	12.92
4	1.34	2.13	2.78	3.75	4.60	7.17	8.61
5	1.30	2.02	2.57	3.36	4.03	5.89	6.87
6	1.27	1.94	2.45	3.14	3.71	5.21	5.96
7	1.25	1.89	2.36	3.00	3.50	4.79	5.41
8	1.24	1.86	2.31	2.90	3.36	4.50	5.04
9	1.23	1.83	2.26	2.82	3.25	4.30	4.78
10	1.22	1.81	2.23	2.76	3.17	4.14	4.59
12	1.21	1.78	2.18	2.68	3.05	3.93	4.32
15	1.20	1.75	2.13	2.60	2.95	3.73	4.07
20	1.18	1.72	2.09	2.53	2.85	3.55	3.85
24	1.18	1.71	2.06	2.49	2.80	3.47	3.75
30	1.17	1.70	2.04	2.46	2.75	3.39	3.65
40	1.17	1.68	2.02	2.42	2.70	3.31	3.55
60	1.16	1.67	2.00	2.39	2.66	3.23	3.46
120	1.16	1.66	1.98	2.36	2.62	3.16	3.37
∞	1.15	1.64	1.96	2.33	2.58	3.09	3.29

of the overall significance of the fit.[3] What about the significance of individual terms? This can be assessed by calculation of the standard error of the regression coefficients, a measure of how much of the dependent variable prediction is contributed by that term. A statistic, the t statistic, may be calculated for each regression coefficient by division of the coefficient by its standard error (SE).

$$t = \left| \frac{b}{\text{S.E. of } b} \right| \tag{6.17}$$

Like the F statistic, the significance of t statistics is assessed by looking up a standard value in a table; the calculated value should exceed the tabulated value. Table 6.5 gives some values of the t statistic for different degrees of freedom and confidence levels. Unlike the F tables, t tables have only one degree of freedom which corresponds to the degree of freedom associated with the error sum of squares. This value is given by

---

[3] As long as the data is well distributed and does not behave as the examples shown in Figure 6.4. This situation is often difficult to check for multivariate data.

$(n-p-1)$, where n is the number of samples in the data set and p is the number of independent variables in the equation, including the constant. It can be seen from the table that the value of t at a 5 % significance level, for a reasonable number of degrees of freedom (five or more), is around two. This is equivalent to saying that the regression coefficient should be at least twice as big as its standard error if it is to be considered significant. Again, this seems like good common sense.

Another useful statistic that can be calculated to characterize the fit of a regression model to a set of data is the standard error of prediction. This gives a measure of how well one might expect to be able to make individual predictions. In the situation where the standard error of measurement of the dependent variable is known, it is instructive to compare these two standard errors. If the standard error of prediction of the regression model is much smaller than the experimental standard error then the model has 'over-fitted' the data, whatever the other statistics of the fit might say. After all, it should not be possible to predict y with greater precision than it was measured, from a model derived from the experimental y values. Conversely, if the prediction standard error is much larger than the experimental standard error, then the model is unlikely to be very useful, although in this case it is likely that the other statistics will also indicate a poor fit. Where the experimental standard error is unknown the standard error of prediction can still be used to assess fit by comparison with the range of measured values. As a rule of thumb, if the prediction standard error is less than 10 % of the range of measurements the model will be useful. For many data sets, particularly from biological experiments, a prediction within 10 % may be regarded as very good. A summary of the statistics that have been described so far is shown in Table 6.6.

Table 6.6   Statistics used to characterize regression equations.

Statistic		Use
Correlation coefficient	r	Gives the direction (sign) and degree (magnitude) of a correlation between two variables
Multiple correlation coefficient	R^2	A measure of how closely a regression model fits a data set
F statistic	F	A measure of the overall significance of a regression model
t statistic	t	A measure of the significance of individual terms in a regression equation
Standard error of prediction	SE	A measure of the precision with which predictions can be made from a regression equation

6.3.1 Creating Multiple Regression Models

Creation of a simple linear regression equation is obvious; there are just two variables involved and all that is required is the estimation of the slope and intercept parameters, usually by OLS. The construction of multiple linear regression equations, on the other hand, is by no means as clear since the selection of independent variables for the equation involves choice. At one time variables were selected for inclusion in a model based on some strong justification. This justification might be some *a priori* idea or hypothesis that the chosen parameters were the best ones for explaining the variance in the dependent variable. Increasingly though, this is not the case and some choice of the independent variables is involved. How can this choice be made? One obvious strategy is to use all of the independent variables and, as long as there is a reasonable[4] number of samples compared with variables in the model, then this can be sufficient. A 'full' regression model like this is in fact the basis of one technique for the creation of multiple regression equations, backward-elimination as discussed later. So, how else can regression models be constructed? The following sections describe the most common methods for producing multiple regression equations.

6.3.1.1 Forward Inclusion

Forward inclusion means, as the name implies, the consideration of each variable in turn for inclusion in the model and then selection of the 'best' variable to produce a 1 term equation. The next step is examination of each of the remaining descriptors and selection of the next best variable to produce a 2 term equation, with the process continuing until none of the remaining variables exceed the inclusion criterion or all of the variables have entered the model. So what is meant by 'best' and what is this inclusion criterion? A commonly used measure is F-to-enter, F_{enter}, which is given by:

$$F_{enter} = \frac{RSS_p - RSS_{p+1}}{RSS_{p+1}/n - p - 1} \qquad (6.18)$$

---

[4] Of course the term 'reasonable' can be contentious but a common rule of thumb is at least 3 times as many data points as terms in the model.

where n is the number of data points, p is the number of terms in the equation and RSS_p and RSS_{p+1} are the residual sums of squares before and after the $(p+1)$th variable is included. The best variable at each step will be the one with the highest value of F-to-enter. The value of F-to-enter can be preset by the user so that descriptors will continue to be added until the F-to-enter values of the remaining variables no longer exceed this value. Once a variable has entered the equation in forward inclusion regression it remains there. In most of the older statistical software packages the default value for F_{enter} is 4 irrespective of the number of variables in the model. This critical F value is based on the upper 5 % point of the F distribution on 1 and $(n - p - 1)$ degrees of freedom which, for $n - p - 1$ of around about 20 or higher is approximately 4 (4.35 in Table 6.4). As an alternative to using F_{enter} values some packages allow the user to specify the corresponding p-values with the selection process stopping when the p-value exceeds some specified level.

An example may serve to illustrate this process. Figure 6.6 shows the output from a forward inclusion regression routine (actually a stepwise routine, see Section 6.3.1.3) applied to a data set of calculated physicochemical properties for a set of antimycin A_1 derivatives; the same compounds mentioned in Chapter 1 (Table 1.1 and Figure 1.7). The output shows that the inclusion criterion is F-to-enter of 4 and the first step has selected melting point as the first variable to enter the regression model. The model already contains a constant and this is the normal procedure when fitting regression models. If theory or some hypothesis dictates that the equation should pass through the origin then the constant term can be eliminated from the fitting process. The next variable to enter the model will be LogPcalc which has an F-to-enter of 11.976. The inclusion process continues for two more steps to produce a final 4 variable model as shown in Figure 6.7. The output shows the number of cases (16), the multiple correlation coefficient, squared multiple correlation coefficient and an adjusted multiple correlation coefficient (see Section 6.4.3) along with the regression coefficients, their standard errors and t-statistics. Fitting a four-term equation to only 16 data points is probably overfitting and a safer model for this set would probably be the three- or even two-term model. This demonstrates one of the dangers of relying blindly on an algorithm to fit models to data. The fitting process has stayed within the constraints of F-to-enter but has produced a model which may not be reliable. The output also reports the analysis of variance sums of squares and an F statistic which may be used to judge the overall fit. This latter quantity may be misleading

```
Dependent Variable ACT
Minimum tolerance for entry into model = 0.000000
Forward stepwise regression with F-to-Enter=4.000 and F-to-Remove=3.900
Step # 1 R =  0.700 R-Square  =  0.490
Term entered: MELT_PT
```

Effect	Coefficient	Std Error	Std Coef	Tol.	df	F	'P'
In							
1 Constant							
22 MELT_PT	0.013	0.004	0.700	1.00000	1	13.453	0.003
Out	Part. Corr.						
2 M_AICH2	-0.169	.	.	0.97220	1	0.382	0.547
3 M_AICH3	0.098	.	.	0.75088	1	0.127	0.727
4 M_AICH4	0.471	.	.	0.99343	1	3.701	0.077
5 M_DIPV_X	0.058	.	.	0.94184	1	0.043	0.838
6 M_DIPV_Y	-0.095	.	.	0.88004	1	0.118	0.737
7 M_DIPV_Z	-0.283	.	.	0.98763	1	1.135	0.306
8 M_DIPMOM	-0.014	.	.	0.92652	1	0.002	0.961
9 M_ESDL5	-0.023	.	.	0.92579	1	0.007	0.935
10 M_ESDL10	-0.190	.	.	0.98889	1	0.489	0.497
11 M_NSDL2	0.223	.	.	0.83566	1	0.683	0.424
12 M_NSDL10	0.199	.	.	0.92309	1	0.535	0.477
13 VDW_VOL	-0.151	.	.	0.60234	1	0.302	0.592
14 M_OF_I_X	0.009	.	.	0.71942	1	0.001	0.975
15 P_E_AX_Y	0.158	.	.	0.95756	1	0.334	0.573
16 MOL_WT	-0.018	.	.	0.96474	1	0.004	0.949
17 S8_1DY	0.036	.	.	0.85695	1	0.017	0.899
18 S8_1DZ	0.334	.	.	0.95276	1	1.632	0.224
19 S8_1CX	-0.012	.	.	0.51974	1	0.002	0.967
20 S8_1CZ	-0.007	.	.	0.79543	1	0.001	0.979
21 LOGPCALC	0.692	.	.	0.90667	1	11.976	0.004
23 SUM_OF_F	0.033	.	.	0.58749	1	0.014	0.907
24 SUM_OF_R	-0.065	.	.	0.93456	1	0.055	0.818

Figure 6.6 Snapshot of the output from the statistics package Systat showing the first step in a forward inclusion regression analysis.

since the model was chosen from a large pool of variables and thus may suffer from 'selection bias' as discussed in Section 6.4.4. Finally, this particular program also reports some problems with individual cases showing that case 8 is an outlier and that case 14 has large leverage. Discussion of these problems is outside the scope of this chapter but can be found in references [3, 4 & 5] and in the help file of most statistics programs.

6.3.1.2   Backward Elimination

This procedure begins by construction of a single linear regression model which contains all of the independent variables and then removes them one at a time. Each term in the equation is examined for its contribution to the model, by comparison of F-to-remove, F_{remove}, for example.

```
Dep Var: ACT    N: 16    Multiple R: 0.930    Squared multiple R: 0.865

Adjusted squared multiple R: 0.816    Standard error of estimate: 0.337

Effect          Coefficient    Std Error    Std Coef Tolerance      t    P(2 Tail)

CONSTANT          -7.487         1.118         0.000       .        -6.695  3.3E-05
M_DIFV_Z          -0.113         0.055        -0.229     0.978      -2.045    0.066
M_ESDL10          -0.879         0.321        -0.318     0.906      -2.736    0.019
LOGPCALC           0.659         0.131         0.607     0.839       5.024  3.8E-04
MELT_PT            0.017         0.002         0.945     0.862       7.920  7.1E-06


Analysis of Variance
Source              Sum-of-Squares    df    Mean-Square      F-ratio          P

Regression              8.001        4        2.000        17.640  9.43691E-05
Residual                1.247       11        0.113


------------------------------------------------------------------------------
*** WARNING ***
Case          8 is an outlier          (Studentized Residual =        -3.000)
Case         14 has large leverage     (Leverage =          0.881)
```

Figure 6.7   Snapshot of the last stage in a forwards inclusion analysis.

F-to-remove is defined by an equation similar to (6.18):

$$F_{remove} = \frac{RSS_{p-1} - RSS_p}{RSS_p / n - p - 1}, \tag{6.19}$$

where the notation is the same as that used in Equation (6.18). The variable making the smallest contribution is removed (lowest F_{remove}) and the regression model is recalculated, now with one term fewer. Any of the usual regression statistics can be used to assess the fit of this new model to the data and the procedure can be continued until a satisfactory multiple regression equation is obtained. Satisfactory here may mean an equation with a desired correlation coefficient or a particular number of independent variables, etc.

Backward elimination and forward inclusion might be viewed as means of producing the same result from opposite directions. However, what may be surprising is that application of the two procedures to the same data set does not necessarily yield the same answer. Newcomers to data analysis may find this disturbing and for some this may reinforce the prejudice that 'statistics will give you any answer that you want', which of course it can. The explanation of the fact that forward inclusion and backward elimination can lead to different models lies in the presence of collinearity and multicollinearity in the data. A multiple regression equation may be viewed as a set of variables which *between* them account for some or all of the variation in the dependent variable. If the independent variables themselves are correlated in pairs (collinearity) or as linear combinations (multicollinearity) then different combinations may account for the same part of the variance of y. An example of this can be seen in Equations (6.20) and (6.21) which describe the pI_{50} for the inhibition of thiopurine methyltransferase by substituted benzoic acids in terms of calculated atomic charges [6].

$$pI_{50} = 12.5q_{2\pi} - 8.3$$
$$n = 15 \quad r = 0.757 \tag{6.20}$$

$$pI_{50} = 12.5q_{6\pi} - 8.4$$
$$n = 15 \quad r = 0.785 \tag{6.21}$$

The individual regression equations between pI_{50} and the two π-electron density parameters have quite reasonable correlation coefficients and thus it might be expected that they would be useful in a multiple regression equation. The two equations, however, are almost identical, indicating a very high collinearity between these descriptors. When combined into a two-term equation (Equation (6.22)), which has an improved correlation coefficient, we see the effect of this collinearity; even the sign of one of the coefficients is changed.

$$pI_{50} = -74q_{2\pi} + 84q_{6\pi} - 6.3$$
$$n = 15 \quad r = 0.855 \tag{6.22}$$

The fact that these two descriptors are explaining a similar part of the variance in the pI_{50} values was revealed in the statistics of the fit for the two-term equation (high standard errors of the regression coefficients). Collinearity and multicollinearity in the descriptor set (independent variables) may lead to poor fit statistics or may cause instability in the regression coefficients. Indeed, regression coefficients which are seen to change markedly as variables are added to or removed from a model are a good indication of the presence of collinear variables.

6.3.1.3   Stepwise Regression

Stepwise regression is a combination of forward inclusion and backward elimination. The first two steps are the same as for forward inclusion and the algorithm selects the two descriptors with the highest F-to-enter values. After this, as each new variable is added a test is performed to see if any of the variables entered at an earlier step can be deleted. The procedure uses both Equations (6.18) and (6.19) in a sequential manner. The stepping stops when no more variables satisfy either the criterion for removal or the criterion for inclusion. To prevent the procedure from unnecessarily cycling the critical values of F-to-enter and F-to-remove should be such that $F_{remove} < F_{enter}$ as can be seen in the third line of the output in Figure 6.6.

All three of these methods used for variable selection are prone to entrapment in local minima, i.e. they find a combination of variables that cannot be improved upon in the next step (removal or addition of one variable) in terms of the criterion function. This can be avoided by performing either a Tabu search (TS) or the more computationally expensive 'all subsets regression' as discussed in the next section. Tabu search is well described by Glover [7, 8].

6.3.1.4   All Subsets

Are there solutions to these problems in the construction of multiple linear regression equations? Does forward inclusion or backward elimination give the best model? Are there alternative ways to construct multiple linear regression equations? One popular approach is to calculate all possible equations of a particular size and then select the best on the basis of the fit statistics. Selection of the best subset of any given size usually involves picking the one which maximizes R^2 although alternative procedures have also been employed. All subset selection techniques will find the combination of variables that maximizes or minimizes a criterion function. This is a property not guaranteed by any of the stepwise methods. At one time this would have been a way of tying up your computer for many hours if not days even for a moderate number of variables from which to select. The speed of modern computers now allows such calculations to be carried out routinely but this procedure will be particularly prone to chance effects (see Section 6.4). There is a practical limit, however, on the size of the models and the pool of variables which can be explored by this means. A pool of 10 variables gives rise to 10 possible 1-term equations, 45 2-term equations, 120 3-term equations and so on. The formula for the number (N) of possible models of size, p, from a pool size, k, is given by:

$$N = k!/(p!(k - p)!) \tag{6.23}$$

Thus, the number of possible models increases very rapidly as the pool size is increased and as the size of the models examined increases. An extreme example, as discussed in the next section, estimates that at a rate of calculation of 1 model per second it would take 226 million years to systematically search for all possible models up to size 29 from a pool of 53 variables!

6.3.1.5 Model Selection by Genetic Algorithm

Darwin's theory of evolution, natural selection, survival of the fittest, selection pressure and so on, appears to offer a neat explanation of the processes leading to the development of highly complex organisms from an enormous variety of options as evidenced by the variety of species on earth. Without getting involved in theological arguments about the origin of species it does appear that this theory proposes a means by which a large solution space – all of nature – can be explored with considerable efficiency. Borrowing from this theory has led to the development of algorithms which make use of these concepts to solve a variety of problems which are difficult if not impossible to solve by exhaustive search of all possible solutions. These approaches are described in more detail in Chapter 9 (Section 9.5) but for the purposes of this discussion suffice it to say that the terms in a regression model may be viewed as chromosomes in an organism.

So, having expressed a multiple linear regression equation as an organism, how is a genetic[5] approach applied? There are some general steps involved:

- Choose a coding scheme for the problem, in this case the variables in a regression model.
- Create an initial population. This is most often carried out by random assignment of variables to the regression models although it can be 'seeded' with known good solutions. Implicit in this, of course, is the question of the size of the models and also the number of equations in the population.
- Choose a fitness function and evaluate the solutions. In this case a fit statistic such as R^2 may be used.

---

[5] This is a generic term which includes genetic algorithms, evolutionary programming and so on. See Chapter 9.

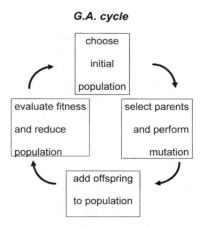

Figure 6.8   Schematic illustration of the genetic process.

- Alter the population by producing new solutions. This may be by mutation of existing solutions or mating (crossover) of two existing solutions.
- Having created the new solutions they are added to the population and the population is then reduced to give a new generation. This reduction is carried out by ranking the solutions in terms of the fitness function and then removing the least fit solutions to give a generation of a set size. In this way existing good solutions are retained in the population.

The genetic process is illustrated in Figure 6.8. The cycle is repeated for a set number of generations or until some target value of the fitness function is achieved. Crucial parameters in this fitting procedure are the size of the population, the nature of the fitness function and the number of cycles that the process runs for, all of which are determined by the user. The data set used to illustrate forward inclusion in Section 6.3.1.1 has been the subject of examination by several genetic techniques. This set consisted of 53 calculated descriptors for a set of 31 compounds. The original report of this study involved the removal of correlated descriptors by application of the CORCHOP algorithm (see Section 3.6 in Chapter 3) to leave 23 variables which was further reduced by the selection of the 10 best variables in terms of their correlation with the dependent variable. All of the genetic approaches used the full set of 53 parameters as potential variables in regression solutions. This is where the 226 million years comes from as mentioned in the previous section. For a set of 31 compounds it is possible to fit models up to a size of 29 variables and from a pool of 53 descriptors this would give rise to 7.16×10^{15} models [9]. In practice, of course, models of such a size would be hopelessly overfitted and it is unlikely that any analysis would consider regression equations larger than say 6 variables for such a data set. In fact, as the original study reported, there was an initial training set of only 16 compounds and so the fitting was restricted to three-term equations.

All of the genetic methods applied to this data set discovered a variety of better models than the original report. This is partly a consequence of the inclusion of all of the original variables in the genetic studies and partly a consequence of the strategy employed in the original study which was a sequential elimination and then selection of variables. This also demonstrates a very important feature of these genetic methods in that they generate a population of possible solutions. All of the methods described above for the production of multiple linear regression models result in a single model at the end of the process, even the all subsets approach only gives a single best model for each chosen subset size. This population of possible solutions allows the generation of a range of predictions and also provides an interesting alternative method for variable selection. The descriptors which emerge in the population of solutions may be ranked in terms of their frequency of occurrence.

6.3.2   Non-linear Regression Models

Non-linear models may be fitted to data sets by the inclusion of functions of physicochemical parameters in a linear regression model – for example, an equation in π and π^2 as shown in Figure 6.5 – or by the use of nonlinear fitting methods. The latter topic is outside the scope of this book but is well covered in many statistical texts (e.g. Draper and Smith [3]). Construction of linear regression models containing non-linear terms is most often prompted when the data is clearly not well fitted by a linear model, e.g. Figure 6.4b, but where regularity in the data suggests that some other model will fit. A very common example in the field of quantitative structure–activity relationships (QSAR) involves non-linear relationships with hydrophobic descriptors such as log P or π. Non-linear dependency of biological properties on these parameters became apparent early in the development of QSAR models and a first

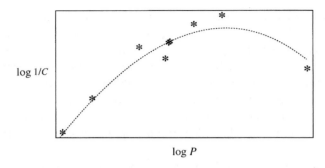

Figure 6.9 Plot of biological response (C is the concentration to give a particular effect) against log P.

approach to the solution of these problems involved fitting a parabola in log P [10].

$$\log {}^1\!/_C = a(\log P)^2 + b \log P + c\sigma + \text{constant} \qquad (6.24)$$

Equation (6.24) may simply contain terms in π or log P, or may contain other parameters such as σ and so on. Unfortunately, many data sets appear to be well fitted by a parabola, as judged by the statistics of the fit, but in fact the data only corresponds to the first half or so of the curve. This is demonstrated in Figure 6.9 for the fit shown in Equation (6.25):

$$\log {}^1\!/_C = 1.37 \log P - 0.35(\log P)^2 + 2.32 \qquad (6.25)$$

Dissatisfaction with the fit of parabolic models such as this and a natural desire to 'explain' QSARs has led to the development of a number of mechanistic models, as discussed by Kubinyi [11]. These models give rise to various expected functional forms for the relationship between biological data and hydrophobicity, and data sets may be found which will be well fitted by them. Whatever the cause of such relationships it is clear that non-linear functions are required in order to model the biological data. An interesting feature of the use of non-linear functions is that it is possible to calculate an optimum value for the physicochemical property involved (usually log P). For example, Equation (6.25) gives an optimum value (at which log $1/C$ is a maximum) for log P of 1.96. This ability to derive optimum values led to attempts to define optima for the transport of compounds across various biological 'barriers'. For example, Hansch and co-workers [12] examined a number of QSARs involving compounds acting on the central nervous system and concluded that the optimum log P value for penetration of the blood–brain barrier was 2. Subsequent work, by Hansch and others, has shown that the prediction of brain uptake is not quite such a simple matter. Van de Waterbeemd and Kansy [13], for example, have demonstrated that brain penetration may be described by a hydrogen-bonding capability parameter (Λ_{alk}) and Van der Waals' volume (V_m).

$$\log(C_{brain}/C_{blood}) = -0.338(\pm 0.03)\Lambda_{alk} + 0.007(\pm 0.001)V_m + 1.73\,(\pm 0.30)$$
$$n = 20, \quad r = 0.934, \quad s = 0.290, \quad F = 58 \qquad (6.26)$$

where the figures in brackets are the standard errors of the regression coefficients and s is the standard error of prediction.

6.3.3   Regression with Indicator Variables

Indicator variables are nominal descriptors (see Chapter 1, Section 1.4.1) which can take one of a limited number of values, usually two. They are used to distinguish between different classes of members of a data set. This situation most commonly arises due to the presence or absence of specific chemical features; for example, an indicator variable might distinguish whether or not compounds contain a hydroxyl group, or have a *meta* substitution. An indicator variable may be used to combine two data sets which are based on different parent structures. Clearly, the dependent data for the different sets should be from the same source, otherwise there would be little point in combining them, and there should be some common physicochemical descriptors (but see later in this section, Free–Wilson method). Indicator variables are treated in multiple regression just as any other variable with regression coefficients computed by least squares. An example of this can be seen in the correlation of reverse phase HPLC capacity factors and calculated octanol/water partition coefficients for the xanthene and thioxanthene derivatives shown in Figure 6.10 [14].

The correlation is given by Equation (6.27) in which the term D was used to indicate the presence ($D = 1$) or absence ($D = 0$) of the –NHCON(NO)– group, in other words series I or series II in Figure 6.10.

$$\log P = 0.813(\pm 0.027)\log k_w + 2.114(\pm 0.161)D \qquad (6.27)$$
$$n = 24 \qquad r = 0.972 \qquad s = 0.365$$

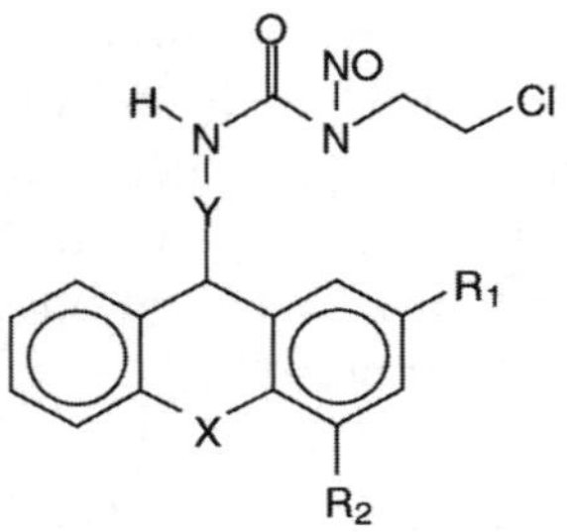

Figure 6.10 Parent structures for the compounds described by Equations (6.27) and (6.28).

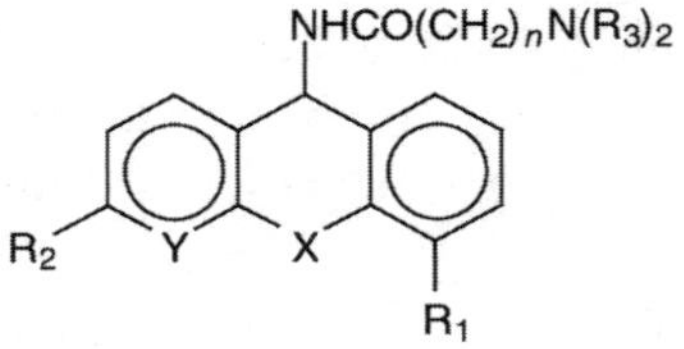

Examination of the log k_w values showed that the replacement of oxygen by sulphur did not produce the expected increase in lipophilicity and it was found that a second indicator variable, S, to show the presence or absence of sulphur could be added to the equation to give:

$$\log P = 0.768(\pm 0.021)\log k_w + 2.115(\pm 0.115)D + 0.415(\pm 0.095)S$$
$$n = 24 \qquad r = 0.985 \qquad s = 0.260 \qquad\qquad (6.28)$$

The correlation coefficient for Equation (6.28) is slightly improved over that for Equation (6.27) (but see Section 6.4.3), the standard error has been reduced, and the regression coefficients for the log k_w and D terms are more or less the same. This demonstrates that this second indicator variable is explaining a different part of the variance in the log P values. It may have been noticed that Equations (6.27) and (6.28) do not contain intercept terms: this is because the intercepts are not significantly different to zero. These examples show how indicator variables can be used to improve the fit of regression models, but do the indicator variables (actually their regression coefficients) have any physicochemical meaning? The answer to this question is a rather unsatisfactory 'yes and no'. The sign of the regression coefficient of an indicator variable shows the direction (to reduce or enhance) of the effect of a particular chemical feature on the dependent variable while the size of the coefficient gives the magnitude of the effect. This does not necessarily bear any relationship to any particular physicochemical property, indeed it may be a mathematical artefact as described later. On the other hand, it may be possible to ascribe some meaning to indicator variable regression coefficients. The log P values used in Equations (6.27) and (6.28) were calculated by the Rekker fragmental method (see Section 9.2.1 and Table 9.2). This procedure relies on the use of fragment values for particular chemical groups and the –NHCON(NO)– group, accounted for by the indicator D, was missing from the scheme. The regression coefficient for this indicator variable has a fairly constant value, 2.114 in Equation (6.27) and 2.115 in Equation (6.28), suggesting that this might be a reasonable estimate for the fragment contribution of this group. Measurement of log P values for two compounds in set I allowed an estimate of $-2.09(\pm0.14)$ to be made for this fragment, in good agreement with the regression coefficient of D. At first sight this statement may seem surprising since the signs of the fragment value and regression coefficients are different. The calculated log P values used in the equations did not take account of the hydrophilic (negative contribution) nitrosureido fragment and thus are bigger, by 2.11, than the experimentally determined HPLC capacity factors.

How does an indicator variable serve to merge two sets of data? The effect is difficult to visualize in multiple dimensions but can be seen in two dimensions in Figure 6.11.

Here, the two lines represent the fit of separate linear regression models, for multiple linear regression these would be surfaces. If the indicator variable has a value of zero for the compounds in set A it will have no effect on the regression line, whatever the value of the fitted regression coefficient. For the compounds in set B, however, the indicator variable has the effect of adding a constant to all the log $1/C$ values (1 $\times$ regression coefficient of the indicator variable). This results in a displacement of the regression line for the B subset of compounds so that it merges with the line for the A subset.

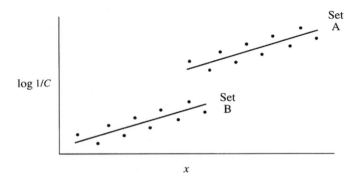

Figure 6.11 Illustration of two subsets of compounds with different (parallel) fitted lines.

An indicator variable can be very useful in combining two subsets of compounds in this way since it allows the creation of a larger set which *may* lead to more reliable predictions. It is also useful to be able to describe the activity of compounds which are operating by a similar mechanism but which have some easily identified chemical differences. However, the situation portrayed in Figure 6.11 is ideal in that the two regression lines are of identical slope and the indicator variable simply serves to displace them. If the lines were of different slopes the indicator may still merge them to produce an apparently good fit to the larger set, but in this case the fitted line would not correspond to a 'correct' fit for either of the two subsets. This situation is easy to see for a simple two-dimensional case but would clearly be difficult to identify for multiple linear regression. A way to ensure that an indicator variable is not producing a spurious, apparently good, fit is to model the two subsets separately and then compare these equations with the equation using the indicator. The situation can become even more complicated when two or more indicator variables are used in multiple regression equations; great care should be taken in the interpretation of such models.

An interesting technique which dates from the early days of modern QSAR, known as the Free and Wilson method [15], represents an extreme case of the use of indicator variables, since regression equations are generated which contain no physicochemical parameters. This technique relies on the following assumptions.

1. There is a constant contribution to activity from the parent structure.
2. Substituents on the parent make a constant contribution (positive or negative) to activity and this is additive.
3. There are no interaction effects between substituents, nor between substituents and the parent.

Of these assumptions, 1 is perhaps the most reasonable and 3 the most unlikely. After all, it is the interaction of substituents with the electronic structure of the parent that gives rise to Hammett σ constants (see Chapter 10). However, despite any misgivings concerning the assumptions,[6] this method has the attractive feature that it is not necessary to

---

[6] The first two assumptions are implicit, although often not stated, in many other QSAR/QSPR methods. The third assumption may be accounted for to some extent by the deliberate inclusion of several examples of each substituent.

Table 6.7 Free–Wilson data table (reproduced from ref. [16] with permission of the Collection of Czechoslovak Chemical Communications).

Compound	R$_1$				R$_2$		
	H	4–CH$_3$	4–Cl	3–Br	H	4'–CH$_3$	4'–OCH$_3$
XI	1	0	0	0	1	0	0
XII	1	0	0	0	0	1	0
XIII	1	0	0	0	0	0	1
XIX	0	1	0	0	0	0	1
XXX	0	0	1	0	0	1	0
XXXV	0	0	0	1	1	0	0

measure or calculate any physicochemical properties; all that is required are measurements of some dependent variable. The technique operates by the generation of a data table consisting of zeroes and ones. An example of such a data set is given in Table 6.7 for six compounds based on the parent structure shown in Figure 6.12.

A Free–Wilson table will also contain a column or columns of dependent (measured) data; for the example shown in Table 6.7 results were given for minimum inhibitory concentration (MIC) against two bacteria, *Mycobacterium tuberculosis* and *Mycobacterium kansasii*. Each column in a Free–Wilson data table, corresponding to a particular substituent at a particular position, is treated as an independent variable. A multiple regression equation is calculated in the usual way between the dependent variable and the independent variables with the regression statistics indicating goodness of fit. The regression coefficients for the independent variables represent the contribution to activity of that substituent at that position, as shown in Table 6.8. In this table, for example, it can be seen that replacement of hydrogen with a methyl substituent (R$_1$) results in a reduction in activity (increase in MIC) against both bacteria.

One of the disadvantages of the Free–Wilson method is that – unlike regression equations based on physicochemical parameters – it cannot

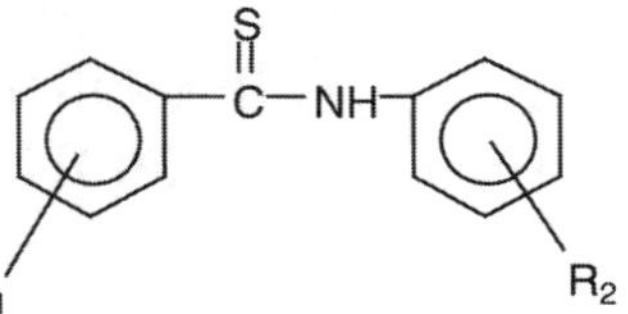

Figure 6.12 Parent structure for the compounds given in Table 6.7 (reproduced from ref. [16] with permission of the Collection of Czechoslovak Chemical Communications).

Table 6.8 Activity contributions for substituents as determined by the Free–Wilson technique (reproduced from ref. [16] with permission of the Collection of Czechoslovak Chemical Communications).

| | Δ MIC against | |
Substituent	*M. kansasii*[a]	*M. tuberculosis*[b]
4–H	−0.397	−0.116
4–CH$_3$	0.264	0.101
4–OCH$_3$	0.290	0.337
4–Cl	0.095	−0.101
3–Br	−0.253	−0.312
4'–H	−0.078	0.088
4'–CH$_3$	0.260	0.303
4'–OCH$_3$	−0.081	0.085
4'–Cl	0.403	0.303
3',4'–Cl$_2$	−0.259	−0.586
4'–C–C$_6$H$_{11}$	−0.589	−0.399
4'–Br	0.345	0.205
μ_o[c]	1.871	1.887

[a]Fit statistics, $r = 0.774$, $s = 0.43$, $F = 3.59$, $n = 35$.
[b]$r = 0.745$, $s = 0.42$, $F = 3.01$, $n = 35$.
[c]μ_o is the (constant) contribution of the parent structure to MIC.

be used to make predictions for substituents not included in the original analysis. The technique may break down when there are linear dependencies between the structural descriptors, for example, when two substituents at two positions always occur together, or where interactions between substituents occur. Advantages of the technique include its ability to handle data sets with a small number of substituents at a large number of positions, a situation not well handled by other analytical methods, and its ability to describe quite unusual substituents since it does not require substituent constant data. A number of variations and improvements have been made to the original Free and Wilson method, these and applications of the technique are discussed in a review by Kubinyi [17].

6.4   MULTIPLE REGRESSION: ROBUSTNESS, CHANCE EFFECTS, THE COMPARISON OF MODELS AND SELECTION BIAS

6.4.1   Robustness (Cross-validation)

The preceding sections have shown how linear regression equations, both simple and multiple, may be fitted to data sets and statistics calculated to characterize their fit. It has also been shown how at least one statistic, the correlation coefficient, can give a misleading impression of how well a regression model fits a data set. This was shown in Figure 6.4 which also demonstrates how easily this may be checked for a simple two-variable problem. A plot of predicted versus observed goes some way towards verification of the fit of multiple regression models but is there any other way that such a fit can be checked? One answer to this problem is a method known as cross-validation or jack-knifing. This involves leaving out a number of samples from the data set, calculating the regression model and then predicting values for the samples which were left out. Cross-validation is not restricted to the examination of regression models; it can be used for the evaluation of any method which makes predictions and, as will be seen in the next chapter, may be used for model selection.

How are the left-out samples chosen? One obvious way to choose these samples is to leave one out at a time (LOO) and at one time this was probably the most commonly used form of cross-validation. Using the LOO method it is possible to calculate a cross-validated R^2, by comparison of predicted values (when the samples were not used to calculate the model) with the measured dependent variable values. This is also sometimes referred to as a prediction R^2, R^2_{cv} or Q^2. Such correlation coefficients will normally be lower than a 'regular' correlation coefficient (but see later) and are said to be more representative of the performance (in terms of prediction) that can be expected from a regression. Other 'predictive' statistics, such as predicted residual sum of squares (PRESS, see Chapter 7), can also be calculated by this procedure. Cross-validation can not only give a measure of the likely performance of a regression model, it can also be used to assess how 'robust' or stable the model is. If the model is generally well fitted to a set of data then omission of one or more sample points should not greatly disturb the regression coefficients. By keeping track of these coefficients as samples are left out, it is possible to evaluate the model for stability, and also to identify which points most affect the fit.

Although LOO cross-validation is the most obvious choice, is it the best? Unfortunately, it is not. Figure 6.13 shows a simple two-dimensional situation in which a straight line model is well fitted to a set of data points which also contains a few outliers.

Some of these points (a and b) will not affect the fitting of the line to the rest of the data and so will be badly predicted (whether included in the model or not) but would not alter the regression coefficients. Other points (c and d) which lie off the line but outside the rest of the data will affect the fit and thus will be badly predicted when left out and will alter the coefficients of the model. So far, so good: LOO cross-validation

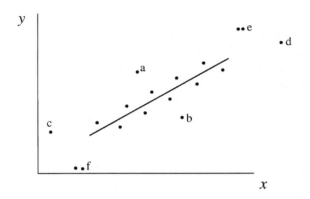

Figure 6.13   Two-dimensional example of a data set with outliers.

would identify these points. Samples e and f, however, occur along with another point well removed from the line and thus LOO would not identify them as being poorly predicted. A solution to this problem is to leave compounds out in groups but the question then arises as to how to choose the groups. The most obvious way to do this is to choose the cross-validation groups at random and to repeat this a sufficient number of times to ensure that the cross-validation results are representative. Cross-validation in groups also results in the need for a lot of computer time to carry out the recalculation of the models and can generate a lot of information which needs to be assessed. This used to be a problem when computers were relatively slow but today they are so fast that the extra computing overhead is rarely a problem. Some statistics packages have cross-validation built in.

The other way to test how well a regression model is fitted to a set of data is to see how well it performs in prediction. Although this isn't really a test of "fit" it can indicate problems in the fitted model. So, in order to test prediction another data set is required as first discussed in chapter 1. This is an evaluation set, that is to say a set of data with results that has been held back from the model fitting process. This is also often called a test set, although this is really incorrect as a true test set is one that becomes available after the model has been fitted, usually because new measurements have been made. As has been pointed out elsewhere there is no standard for the nomenclature used in naming these sets; many of the test sets shown in the examples discussed in chapter 4, for instance, are actually evaluation sets. Perhaps because of this lack of standardization, the term Q^2 is sometimes used to refer to the prediction results for an evaluation or test set, when really it should be R^2_{eval} or R^2_{test}. Bearing this in mind it is important to read literature reports of model fitting and evaluation carefully in order to discern what the reported statistics actually mean.

A problem with the use of evaluation or test sets arises with the question of which mean should be used to calculate the sums of squares necessary to give the correlation coefficient. An OECD guideline (OECD Document ENV/JM/MONO(2007)2, 2007, pp 55) oddly suggests the use of the original training set mean. Schüürmann and colleagues have pointed out that there can be problems with using the training set mean including the contradictory result of obtaining Q^2 values higher than the fitted R^2 results [18]. The recommendation from this study is that the training set mean should be used when calculating cross-validation statistics, either in groups or LOO, but that the test set mean should be used when computing statistics for a test set.

Cross-validation is a useful technique for the assessment of fit and predictive performance of regression (and other) models but it is not the perfect measure that it was once proposed to be, particularly LOO cross-validation. A good solution to the questions of robustness and predictive performance is to use well-selected training and test sets, but this is a luxury we cannot always afford.

6.4.2   Chance Effects

One of the problems with regression analysis, and other supervised learning methods, is that they seek to fit a model. This may seem like a curious statement to make, to criticize a method for doing just what it is intended to do. The reason that this is a problem is that given sufficient opportunity to fit a model then regression analysis will find an equation to fit a data set. What is meant by 'sufficient opportunity'? It has been proposed [19] that the greater the number of physicochemical properties that are tried in a regression model then the greater the likelihood that a fit will be found by chance. In other words, the probability of finding a chance correlation (not a true correlation but a coincidence) increases as the number of descriptors examined is increased. Will not the statistics of the regression analysis fit indicate such an effect? Unfortunately, the answer is no; a chance correlation has the same properties as a true correlation and will appear to give just as good (or bad) a fit.

Do such chance correlations happen, and if so can we guard against them? The fact that they do occur has been confirmed by experiments involving random numbers [20]. Sets of random numbers were generated, one set chosen as a dependent variable and several other sets as independent variables, and the dependent fitted to the independents

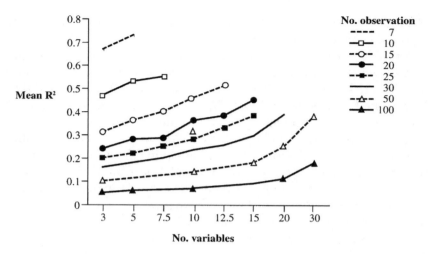

Figure 6.14   Plot of mean R^2 versus the number of screened variables for regression equations generated for sets of random numbers (reproduced from ref. [20] with permission of the American Chemical Society).

using multiple regression. This procedure was repeated many times and occasionally a 'significant' correlation was found. A plot of average R^2 versus the number of random variables screened, for data sets containing different numbers of samples, is shown in Figure 6.14. As originally proposed, the probability of finding a chance correlation for a given size of data set increases as the number of screened variables is increased. Plots such as that shown in the figure may be used to limit the number of variables examined in a regression study although it should be pointed out that these results apply to random numbers, and real data might be expected to behave differently.

Another way in which we can assess the possibility of chance correlations is a method known as Y scrambling. This is performed by randomly swapping the values of the response variable around while maintaining the structure of the independent variables. The dependent variable thus has the same distribution as the original data, mean, standard deviation and so on, but as the values have been scrambled any significant relationships with the independent variables should have been destroyed. Having scrambled the Y values, the regression model is recalculated, the Y values are scrambled again and the regression model recalculated and this is carried out a number of times. The results for the scrambled sets are collected and their correlation coefficients compared with the correlation coefficient of the original model. Figure 6.15 shows the results of Y scrambling 500 times for a 2 term regression model computed on 25

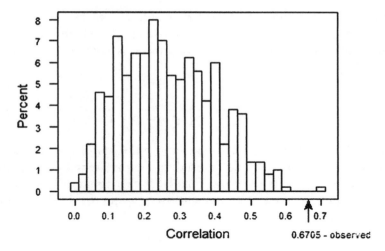

Figure 6.15 Plot of frequency of correlation coefficient values for 500 Y scramblings of a regression model.

data points. The correlation coefficient of the original regression model was 0.67 and as can be seen from the plot all the scrambled correlation coefficients except 1 are lower than the original model. Thus we can be quite confident that the original model is not a chance fit. Having said all this, perhaps the best test of the significance of a regression model is how well it performs with a real set of test data.

6.4.3   Comparison of Regression Models

If two regression equations contain the same number of terms and have been fitted to the same number of data points, then comparison is simple. The R^2 values will show which equation has the best fit and the F and t statistics may be used to judge the overall significance and the significance of individual terms. Obviously, if one equation is to be preferred over another it is expected that it will have significant fit statistics. Other factors may influence the choice of regression models, such as the availability or ease of calculation of the physicochemical descriptors involved. If the regression equations involve different numbers of terms (independent variables), then direct comparison of their correlation coefficients is not meaningful. Since the numerator for the expression defining the multiple correlation coefficient (Equation (6.14)) is the explained sum of squares, it is to be expected that this will increase as extra terms are added to a regression model. Thus, R^2 would be expected to increase as extra terms are added. An alternative statistic to R^2, which takes account of the number of terms in a regression model, is known as the adjusted R^2 coefficient ($\overline{R}^2$)

$$\overline{R}^2 = 1 - (1 - R^2)\frac{n-1}{n-p-1} \qquad (6.29)$$

where n is the number of data points and p the number of terms in the equation. This statistic should be used to compare regression equations with different numbers of terms. Finally, if two equations are fitted to different numbers of data points, selection of a model depends on how it is to be used.

6.4.4 Selection Bias

The selection of variables for inclusion in a multiple regression model or, for that matter, many kinds of mathematical models describing the relationship between a response variable and descriptor variables is often a very necessary step. There are situations where variable selection is not necessary, when testing an hypothesis for example or where the analytical method is able to cope with extra unneeded variables, but more often than not some selection process is required. Apart from the problem of deciding which variables to choose, are there any other drawbacks to variable selection? The unfortunate answer to this question is yes. When variables are selected from a large pool of potential variables they are usually chosen to maximize some function involving the response variable, such as R^2. This selection process suffers from an effect known as competition or selection bias. A simple experiment with random numbers will demonstrate this:

- Four sets of 25 random numbers were generated to represent a dependent (y) variable and three independents ($x1$, $x2$, $x3$). These were standard normal variables, that is to say with zero mean and unit variance.
- One x variable was chosen at random and the y variable regressed on it and the regression coefficient and R^2 values recorded.
- Step 2 was repeated but the y variable was regressed on all three x variables and the result with the highest correlation was recorded.
- The whole process was repeated a 1000 times.

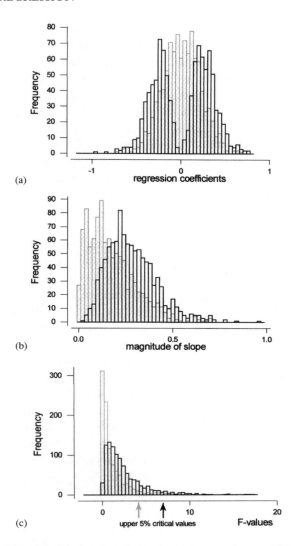

Figure 6.16 Plot of frequency of occurrence of (a) regression coefficients (b) magnitude of regression coefficients and (c) F values for randomly selected (grey) regression models and models selected to maximize R^2 (black) (reproduced from ref. [21] with permission of Wiley-VCH)

Figure 6.16 shows the results of this experiment where the values obtained from random selection are shown in grey and those from maximum R^2 are shown in black.

The first part of the figure shows the histograms for the regression coefficients which, for the randomly chosen x, are fairly normal in shape and centred around zero. Since the variables are random numbers we would expect no correlation between them and the regression coefficient (m in $y = mx + c$) should be zero. The result for the maximum correlation set are quite different, however, and here we see a bimodal distribution which shows that the regression coefficients are centred about two values (negative and positive) which are smaller (larger) than zero. In the second part of the figure where the absolute values of the regression coefficients are plotted this result is even clearer. So, this is one effect of selection bias; the regression coefficients are larger in magnitude than those chosen at random. There is, however, another perhaps more alarming effect of selection bias and that is that it has a marked effect on significance levels. This can be seen in the third part of the figure where the distribution of F ratios for the different models are plotted. The grey arrow marked on the figure is the value of F (4.35) for the 95th percentile from the 1000 simulations for the random selections, that is to say the value of F for the 950th regression when the 1000 values are arranged in ascending numerical order. Because the experiment is fitting y to one x with a sample size of 25 this should correspond with the upper 5 % point of the F-distribution on 1 and $25 - 1 - 1 = 23$ degrees of freedom, i.e. 4.30. The two values are in good agreement which shows that in the case of this simulation 95 % of the regression models would have an F value which doesn't exceed the critical value and thus would be deemed to be non significant. If we now look at the F distribution from the regressions chosen to maximize the fit we find that over 15 % of the values exceed this critical value thus, using the tabulated value for F, it would be concluded that the significance level is 15 % not 5 %. To assess the significance of a model selected when maximizing R^2 would require an F value of approximately 7.1 which is the 95th percentile of the F values from maximum fit.

These F values have been termed F_{max} and simulations have been run using various combinations of sample size (n), model size (p) and pool size (k) to provide a more appropriate set of critical values to judge the significance of multiple linear regression models constructed from large pools of variables [22, 23]. The antimycin analogue data set used as an example in forward inclusion regression (Section 6.3.1.1) and genetic algorithm selection (6.3.1.5) serves well to illustrate the use of these F_{max} values. The first line of Table 6.9 shows the F values obtained from fitting regression models containing 1, 2 and 3 terms and the second row gives the usual F values from tables.

Based on this evidence it would appear that all three models are highly significant. The third line of the table shows the F_{max} values obtained

Table 6.9 Comparison of actual, tabulated, and simulated F-values for the antimycin data (reproduced from ref. [22] with permission of the American Chemical Society (2005)).

Terms in the model	1	2	3
F value from fit	13.55	18.0	17.5
Tabulated F	4.62	3.85	3.54
F_{max} (k)	4.58	3.63	3.48
F_{max} (10)	11.06	10.03	9.80
F_{max} (23)	13.88	14.85	17.39
F_{max} (53)	17.22	21.11	29.73

from simulations in which the pool size was equal to the size of the fitted model $(k = p)$ and these of course should correspond to the tabulated F values which, within the limits of precision of the simulation process, they do. The fourth line of the table shows F_{max} values obtained from simulations with a pool size of 10 variables, the number of variables used in the original reported stepwise regression, and here we can see that the F_{max} values are considerably larger. On the basis of these results the original models would still be judged as 'significant' but it is plain that the regular tabulated values are inappropriate. The last two rows of the table give the F_{max} values obtained from simulations with pool sizes of 23 (the variables used to choose the subset of 10 from) and 53 which was the original starting pool of variables before unsupervised variable elimination. It is left to the reader to judge the significance of these regression models.

Tabulations of F_{max} from these simulation experiments are useful but of course in practice tabulations often lack the precise combination of n, p and k which is needed to assess a particular regression model. It is possible to estimate F_{max} values by means of a power function which was fitted to the simulation results and this estimator is available on the website of the Centre for Molecular Design at the University of Portsmouth (http://www.port.ac.uk/research/cmd/research/selectionbiasinmultiple regression/) as is access to the F_{max} simulator.

6.5 SUMMARY

Regression analysis is a very useful tool for the identification and exploitation of quantitative relationships. The fit of regression models may be readily estimated and the direction and magnitude of individual correlations may give some useful clues as to mechanism. Regression models are easily interpreted, since they mimic a natural process by which we try to relate cause and effect, but it should be remembered that a correlation does not prove such a relationship. Successful regression models may inspire us to design experiments to examine causal relationships and, of course, empirical predictions are always of use. There are dangers in the use of regression analysis – even quite simple models may be very misleading if judged by their statistics alone – but there are means by which some of the dangers may be guarded against. This chapter has been a brief introduction to some of the fundamentals of regression analysis; for further reading see Draper and Smith [3], Montgomery and Peck [4], or Rawlings [5].

In this chapter the following points were covered:

1. how simple linear regression works and how to judge the quality of the models;
2. multiple linear regression models – what they are and how to construct them;
3. how to judge the quality and performance of multiple linear regression models;
4. non-linear regression modelling;
5. chance effects and how to avoid them;
6. the problems of selection bias and how it inflates regression coefficients and the statistics used to judge them;
7. regression models with indicator variables – what they are and how to interpret them.

REFERENCES

[1] Anscombe, F.J. (1973). *American Statistician*, **27**, 17–21.
[2] Lindley, D.V. and Miller, J.C.P. (1953). *Cambridge Elementary Statistical Tables*. Cambridge University Press.
[3] Draper, N.R. and Smith, H. (1981). *Applied Regression Analysis*. John Wiley & Sons, Inc., New York.
[4] Montgomery, D.C. and Peck, E.A. (1982). *Introduction to Linear Regression Analysis*. John Wiley & Sons, Inc., New York.
[5] Rawlings, J.O. (1988). *Applied Regression Analysis; A Research Tool*. Wadsworth & Brooks, California.
[6] Kikuchi, O. (1987). *Quantitative Structure–Activity Relationships*, **6**, 179–84.
[7] Glover, F. (1989). *ORSA Journal on Computing*, **1**, 190–206.
[8] Glover, F. (1990). *ORSA Journal on Computing*, **2**, 4–32.
[9] Kubinyi, H. (1994). *Quantitative Structure–Activity Relationships*, **13**, 285–94.

[10] Hansch, C., Steward, A.R., Anderson, S.M., and Bentley, J. (1968). *Journal of Medicinal Chemistry*, 11, 1–11.

[11] Kubinyi, H. (1993). *QSAR: Hansch Analysis and Related Approaches*, Vol. 1 of *Methods and Principles in Medicinal Chemistry*, (ed. R. Mannold, P. Krogsgaard-Larsen, H. Timmerman), pp. 68–77. VCH, Weinheim.

[12] Hansch, C. (1969). *Accounts of Chemical Research*, 2, 232–9.

[13] Van de Waterbeemd, H. and Kansy, M. (1992). *Chimia*, 46, 299–303.

[14] Fillipatos, E., Tsantili-Kakoulidou, A., and Papadaki-Valiirake, A. (1993). In *Trends in QSAR and Molecular Modelling 92* (ed. C.G. Wermuth), pp. 367–9. ESCOM, Leiden.

[15] Free, S.M. and Wilson, J.W. (1964). *Journal of Medicinal Chemistry*, 7, 395–9.

[16] Waisser, K., Kubicova, L., and Odlerova, Z. (1993). *Collection of Czechoslovak Chemical Communications*, 58, 205–12.

[17] Kubinyi, H. (1988). *Quantitative Structure–Activity Relationships*, 7, 121–33.

[18] Schuurmann, G., Ebert, R.-U., Chen, J., Wang, B. and Kuhne, R. (2008). *J. Chem. Inf. Model.*, 48, 2140–5.

[19] Topliss, J.G. and Costello, R.J. (1972). *Journal of Medicinal Chemistry*, 15, 1066–8.

[20] Topliss, J.G. and Edwards, R.P. (1979). *Journal of Medicinal Chemistry*, 22, 1238–44.

[21] Livingstone, D.J. and Salt, D.W. (2005) Variable selection – spoilt for choice? in *Reviews in Computational Chemistry*, Vol 21, K. Lipkowitz, R. Larter and T.R. Cundari (eds), pp. 287–348, Wiley-VCH.

[22] Livingstone, D.J. and Salt, D.W. (2005). *Journal of Medicinal Chemistry*, 48, 661–3.

[23] Salt, D.W., Ajamani, S., Crichton, R. and Livingstone, D.J. (2007). *Journal of Chemical Information and Modeling*, 47, 143–9.

7

Supervised Learning

Points covered in this chapter

- Supervised learning methods for classified data – discriminant analysis and SIMCA
- Regression on principal components
- Partial Least Squares regression (PLS)
- Continuum regression
- Feature selection

7.1 INTRODUCTION

The common feature underlying supervised learning methods is the use of the property of interest, the dependent variable, to build models and select variables. Regression analysis, which warranted a chapter of its own because of its widespread use, is a supervised learning technique. Supervised methods are subject to the danger of chance effects (as outlined in Section 6.4.2 for regression) which should be borne in mind when applying them. The dependent variable may be classified, as used in discriminant analysis described in the first section of this chapter, or continuous. Section 7.3 discusses variants of regression, which make use of linear combinations of the independent variables; Section 7.4 describes supervised learning procedures for feature selection.

A Practical Guide to Scientific Data Analysis   David Livingstone
© 2009 John Wiley & Sons, Ltd

7.2 DISCRIMINANT TECHNIQUES

The first two parts of this section describe supervised learning methods which may be used for the analysis of classified data. There are a number of techniques which can be used to build models for such data and one has already been discussed, k-nearest-neighbours, but here we will consider only two: discriminant analysis and SIMCA. The use of artificial neural networks to analyse classified data is described in Chapter 9 (Section 9.3.2) and consensus models which use several classification methods are discussed in Section 9.6. Discriminant analysis is related to regression while the other technique, SIMCA, has similarities with principal component analysis (PCA). The final parts of this section describe confusion matrices and consider some of the conditions which data should meet when analysed by discriminant techniques.

7.2.1 Discriminant Analysis

Discriminant analysis, also known as the linear learning machine,[1] is intended for use with classified dependent data. The data may be measured on a nominal scale (yes/no, active/inactive, toxic/nontoxic) or an ordinal scale (1,2,3,4; active, medium, inactive) or may be derived from continuous data by some rule (such as 'low' if <10, 'high' if >10). The objective of regression analysis is to fit a line or a surface *through* a set of data points; discriminant analysis may be thought of as an orthogonal process to this in which a line or surface is fitted in between two classes of points in a data set. This is illustrated in Figure 7.1 where the points represent compounds belonging to one of two classes, A or B, and the line represents a discriminating surface.

It is confusing, perhaps, that the discriminant function itself (shown by the dotted line in Figure 7.1) does run through the data points; the discriminant surface represents some critical value of the discriminant function, often zero. Projection of the sample points onto this discriminant function yields a value for each sample and classification is made by comparison of this value with the critical value for the function. In this simple two-dimensional example, the discriminant function is a straight line; in the case of a set of samples described by P physicochemical

---

[1] *Linear* discriminant analysis is equivalent to the linear learning machine. There are also procedures for non-linear discriminant analysis (as there are for non-linear regression) but these will not be considered here.

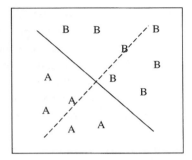

Figure 7.1 Two-dimensional representation of discriminant analysis. The dotted line represents the discriminant function and the solid line a discriminant surface which separates the two classes of samples.

properties, the discriminant function would be a P-dimensional hypersurface. The discriminant function may be represented as

$$W = a_1x_1 + a_2x_2 + \ldots\ldots a_px_p \tag{7.1}$$

or more succinctly as

$$W = \sum_{i=1}^{P} a_ix_i \tag{7.2}$$

where the x_i's are the independent variables used to describe the samples and the a_i's are fitted coefficients. These coefficients are known as the discriminant weights and may be rescaled to give discriminant loadings [1] which are the loadings of the variables onto the discriminant function, reminiscent of principal component loadings, and which are in fact the simple correlation of each variable with the discriminant function. Two things may be noticed from this equation and the figure. The combination of variables is a linear combination, thus this method strictly should be called linear discriminant analysis (LDA). The line shown drawn in the figure (which is at right angles to the discriminant function) is not the only line that could be drawn between the two classes of compounds. Creation of a different discriminant surface (drawing another line) is achieved by computing a different discriminant function. Unlike regression analysis, where the least squares estimate of regression coefficients can give only one answer, the coefficients of the variables in a discriminant function may take one of a number of values. As long as the discriminant function correctly classifies all of the samples, then it does not matter what the values of the discriminant loadings are. This is not to say that an individual discriminant analysis procedure run repeatedly on the same data set will give different answers, but that discriminant functions are more susceptible to change when samples or descriptors are added to or removed from the data set. The fact that a unique solution does not exist for a discriminant function has implications in the use of discriminant analysis for variable selection, as described in Section 7.4. Figure 7.1 also demonstrates how one of the alternative names, the linear learning machine, arose in artificial intelligence research. The algorithm to generate the discriminant function is a 'learning machine' which aims to separate the two classes of samples in a linear fashion by 'learning' from the data.

Returning to Equation (7.1), how is this used? Once a discriminant function has been generated, a prediction can be made for a compound by multiplication of the descriptor variables by their coefficients and summation of these products. This will give a value of W which, if it exceeds some critical value, will assign the compound to one of the two classes. A value of zero may be used as the critical value so that if W is positive the compound belongs to class 1 (A in Figure 7.1), if negative then to class 2 (B in Figure 7.1). The question also arises of how to judge the quality of a discriminant function. In terms of prediction, this is quite easily done by comparison of the predicted class membership with known class membership. A more stringent test of the predictive ability of discriminant analysis is to use a leave one out (LOO) cross-validation procedure as described for regression analysis in Section 6.4.1. It is also possible to compute a statistic for discriminant analysis which is equivalent to the F statistic used to characterize the fit of regression models. This statistic may be used to judge the overall significance of a discriminant analysis result and, in a slightly different form as a partial F statistic, also used to construct discriminant functions (see ref. [1]).

An early example of the use of discriminant analysis in QSAR involved inhibition of the enzyme monoamine oxidase (MAO) by derivatives of aminotetralins and aminoindans shown in Figure 7.2 ([2], see also Section 5.5).

These compounds inhibited the enzyme *in vitro*, and it was possible to obtain percentage inhibition data for them in an enzyme assay but the crucial test was a measure of their activity *in vivo*. This was assessed by judgement of the severity of symptoms following administration of *dl*-Dopa and was given a score of 0, 1, 2, or 3 as shown in Table 7.1. Initial examination of the data by discriminant analysis suggested that

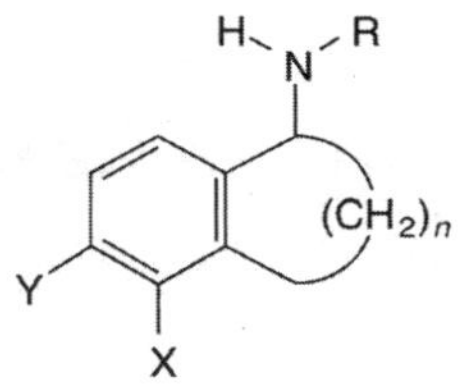

Figure 7.2 Parent structure for the compounds shown in Table 7.1 (reproduced from ref. [2] copyright American Chemical Society).

the compounds should be classified into just two groups,[2] and thus a two-group discriminant function was fitted as shown in the table. This function involved the steric parameter, E_s^c, and an indicator variable which showed whether the compounds were substituted in position X or Y in Figure 7.2. The results of classification by this function are quite impressive: only one compound (number 18) is misclassified. A more exacting test of the utility of discriminant analysis was carried out by fitting a discriminant function to 11 of the compounds in Table 7.1 (indicated by a *). Once again, E_s^c and the indicator variable were found to be important; one of the training set compounds was misclassified and all of the test set compounds were correctly assigned.

A comparison of the performance of discriminant analysis and other analytical techniques on the characterization of two species of ants by gas chromatography was reported by Brill and co-workers [3]. Samples of two species of fire ants, *Solenopsis invicta* and *S. richteri*, were prepared using a dynamic headspace analysis procedure. The gas chromatography analysis resulted in the generation of 52 features, retention data for cuticular hydrocarbons, for each of the samples. Each of these descriptors was examined for its ability to discriminate between the two species of ants and the three best features selected for use by the different analytical methods. A nonlinear map of the samples described by these three features is shown in Figure 7.3 where it is clear that the two species of ants are well separated.

Analysis of this data was carried out by k-nearest-neighbour (see Section 5.1), discriminant analysis and SIMCA (see Section 7.2.2). The discriminant analysis routine used was the linear learning machine (LLM) procedure in the pattern recognition package ARTHUR. A training/test set protocol was used in which the data was split up into two sets five

---

[2] This was supported by the pharmacology since on retest, some compounds moved between groups 0 and 1 or 2 and 3, but rarely from (0,1) to (2,3) or vice versa.

Table 7.1 Structure, physicochemical properties, and potency as MAO inhibitors (in vivo) of the aminotetralins and aminoindans shown in Figure 7.2 (reproduced from ref. [2] with permission of the American Chemical Society).

Number	Structure				Properties[a]		Potency		
	n[b]	R	X	Y	Π	E_s[c]	Observed[c]	Calculated[d]	Calculated[e]
1	2	CH$_3$	H	OCH$_3$	1.3	0.00	3	1	1*
2	3	H	OCH$_3$	H	1.2	0.32	3	1	0*
3	3	H	H	OCH$_3$	1.3	0.32	3	1	1
4	3	CH$_2$CH$_3$	H	OCH$_3$	2.2	−0.07	3	1	1
5	3	CH$_3$	H	OCH$_3$	1.7	0.00	2	1	1*
6	3	CH$_3$	H	OH	1.7	0.00	2	1	1
7	2	H	H	OCH$_3$	0.8	0.32	2	1	1*
8	3	CH$_3$	OCH$_3$	H	1.7	0.00	1	1	1*
9	3	(CH$_2$)$_2$OCH$_3$	H	OCH$_3$	1.7	−0.66	1	0	0*
10	3	(CH$_2$)$_2$CH$_3$	H	OCH$_3$	2.7	−0.66	1	0	0*
11	3	(CH$_2$)$_5$CH$_3$	H	OCH$_3$	4.2	−0.68	1	0	0
12	3	CH$_2$C$_6$H$_5$	OCH$_3$	H	3.5	−0.68	1	0	0
13	3	(CH$_2$)$_2$OH	H	OCH$_3$	1.0	−0.66	1	0	0
14	3	CH$_3$	OH	H	1.7	0.00	0	0	0*
15	3	CH(CH$_3$)$_2$	OCH$_3$	H	2.6	−1.08	0	0	0
16	3	CH(CH$_3$)$_2$	H	OCH$_3$	2.6	−1.08	0	0	0*
17	2	CH(CH$_3$)$_2$	H	OCH$_3$	2.1	−1.08	0	0	0
18	2	H	OCH$_3$	H	0.8	0.32	0	1	0*
19	3	(CH$_2$)$_2$CH$_3$	H	OCH$_3$	1.4	−0.66	0	0	0*
20	3	(CH$_2$)$_6$CH$_3$	H	OCH$_3$	4.7	−0.68	0	0	0

[a] Two indicator variables were also used to distinguish indans and tetralins and the position of substitution (X or Y).
[b] Where n is the n in Figure 7.2.
[c] Activity was scored (0–3) according to the severity of symptoms; scores 2 and 3 are active compounds, 0 and 1 inactive.
[d] Calculated from a two-group discriminant function.
[e] Calculated from a two-group discriminant function which was trained on half of the compound set (indicated by *).

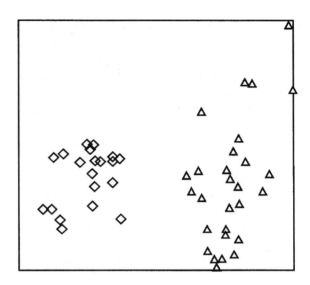

Figure 7.3 Nonlinear map of two species of fire ant, *Solenopsis invicta* (◊) and *Solenopsis richteri* (Δ), described by three GC peaks (reproduced from ref. [3] with permission of Elsevier).

times, with all of the original data used as a test set member at least once. Results of these analyses are shown in Table 7.2. These rather impressive results are perhaps not surprising given the clear separation of the two species shown in Figure 7.3. They do, however, illustrate an important general feature of data analysis and that is that there is no 'right' way to analyse a particular data set. As long as the method used is suitable for the data, in this case classified dependent data, then comparable results should be obtained if the data contains appropriate information.

Table 7.2 Classification of fire ants characterized by gas chromatography data (reproduced from ref. [3] with permission from Elsevier).

Method	Category*	Training correct (%)	Test correct (%)	Overall (%)
KNN	1	100	100	100
($k = 10$)	2	100	100	100
LLM	1	100	89.6	97.9
	2	100	100	100
SIMCA	1	100	100	100
	2	100	100	100

*1 = S. richteri, 2 = S. invicta.

These two examples have involved classified dependent data and continuous independent data. It is also possible to use classified independent data in discriminant analysis, as it is in regression (such as Free–Wilson analysis) or a combination of classified and continuous independent variables. Zalewski [4] has reported a discriminant analysis of sweet and nonsweet cyclohexene aldoximes using indicator variables describing chemical structural features. From the discriminant function it was possible to identify features associated with the two classes of compounds:

Sweet   a short chain up to three carbons
           a substituted carbon at a particular position
           a cyclohexane ring
Nonsweet   a carbon chain at a different position to the sweet compounds
           the presence of a heteroatom in a ring

This function classified correctly 22 out of 23 sweet compounds and 24 out of 29 nonsweet derivatives. The same report also described a discriminant analysis of another set of aldoxime derivatives characterized by molecular connectivity indices (see Chapter 10). The discriminant function involved just two of these indices [5]

$$D_F = 1.21 \, {}^1\chi - 3.88 \, {}^4\chi_p \tag{7.3}$$

The first term $({}^1\chi)$ in Equation (7.3) describes the size of molecules in terms of the number of bonds, the second term $({}^4\chi_p)$ is influenced by the size of the substituents. Compounds with a value of D_F greater than -3.27 were classified as sweet. This discriminant function correctly assigned nine out of ten sweet compounds and eight out of ten nonsweet.

How are discriminant functions built? The construction of a multiple variable discriminant function presents similar problems to the construction of multiple regression models (see Section 6.3.1) and similar solutions have been adopted. The stepwise construction of discriminant functions presents an extra problem in that there is not necessarily a unique solution, unlike the situation for regression. The other question that may have occurred to a reader by now is whether discriminant analysis is able to handle more than two classes of samples. The answer is yes, although the number of discriminant functions needed is not known in advance. If the classes are organized in a uniform way in the multidimensional space, then it may be possible to classify them

Table 7.3 Classification of cancer patients based on tumour marker measurements (reproduced from ref. [6] with permission of John Wiley & Sons, Ltd).

Discrimination between	LDA correct (%)[a]		KNN (%)
	Classification	Prediction	
Controls - patients[b]	83.1	83.7	85.7
Controls - lungs	92.8	91.2	86.8
Controls - breast	79.1	81.1	81.5
Controls - gastro	93.4	91.9	89.1

[a]The data from 102 subjects (30 controls, 72 patients) was divided into training set (classification) and test set (prediction) containing about 70 % and 30 % of the subjects respectively.
[b]The difference between controls and all patients.

using a single discriminant function, e.g. if W (from Equation 7.1) $< x$ then class 1, if $x < W < y$ then class 2, if $W > y$ then class 3 (where x and y are some numerical limits). When the classes are not organized in such a convenient way, then it will be necessary to calculate extra discriminant functions; in general, k classes will require $k - 1$ discriminant functions.

An example of multi-category discriminant analysis can be seen in a report of the use of four different tumour markers to distinguish between controls (healthy blood donors) and patients with one of three different types of cancer [6]. The results of this analysis are shown in Table 7.3 where LDA is compared with k-nearest-neighbour (KNN, k was not specified).

7.2.2 SIMCA

The SIMCA[3] method is based on the construction of principal components (PCs) which effectively fit a box (or hyper-box in P dimensions) around each class of samples in a data set. This is an interesting application of PCA, an unsupervised learning method, to different classes of data resulting in a supervised learning technique. The relationship between SIMCA and PLS, another supervised principal components method, can be seen clearly by reference to Section 7.3.

---

[3] The meaning of these initials is variously ascribed to Soft Independent Modelling of Class Analogy, or Statistical Isolinear MultiCategory Analysis, or SImple Modelling of Class Analogy.

How does SIMCA work? The steps involved in a SIMCA analysis are as follows:

1. The data matrix is split up into subsets corresponding to each class and PCA is carried out on the subsets.
2. The number of PCs necessary to model each class is determined (this effectively defines the hyper-box for each category).
3. The original descriptors are examined for their discriminatory power and modelling power (see below) and irrelevant ones discarded.
4. PCA is again carried out on the reduced data matrix and steps 2 and 3 repeated. This procedure continues until consistent models are achieved.

The discriminatory power and modelling power parameters are measures of how well a particular physicochemical descriptor contributes to the PCs in terms of the separation of classes and the position of samples within the classes. Since the PCs are recalculated when descriptors with low discriminatory or modelling power are removed, new parameters must be recalculated for the remaining properties in the set. Thus, the whole SIMCA analysis becomes an iterative procedure to obtain an optimum solution. Readers interested in further mathematical details of the SIMCA method should consult the chapter by Wold and Sjostrom [7].

The results of applying the SIMCA procedure can perhaps best be seen in a diagram such as that shown in Figure 7.4. The hyper-boxes

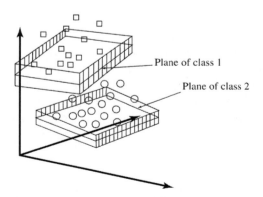

Figure 7.4 Graphical representation of SIMCA (reproduced from ref. [8] with permission of the American Chemical Society).

Table 7.4 Comparison of SIMCA and KNN for classification of water samples (reproduced from ref. [9] with permission from Energia Nuclear & Agricultura).

Region	Number of samples	Number of points incorrectly classified	
		9-NN	SIMCA
Serra Negra	46	2	16
Lindoya	24	2	5
Sao Jorge	7	1	0
Valinhos	39	2	3
Correct (%)		93.8	79.3

do not fit around all of the points in each class but for the purposes of prediction it is possible to assign a sample to the nearest hyper-box. The size and shape of the hyper-boxes allows a probability of class membership to be assigned to predictions and if the objects within a class have some associated continuous property, it is possible to make quantitative predictions (by the position of a sample within the hyper-box compared to other points).

The SIMCA technique has been applied to a variety of problems within the QSAR field and others. One data set that has already been cited (Section 5.2) was also analysed by SIMCA [9]. This data consisted of water samples characterized by their concentrations of four elements (the four most important for classification were chosen from a total of 18 elements measured). A comparison of the performance of SIMCA and the worst nearest neighbour analysis is shown in Table 7.4. For a test set of seven samples, four Lindoya and three Serra Negra, the 9-NN analysis classified all correctly while the SIMCA method misclassified two of the Serra Negra samples as Valinhos.

Another example of the use of SIMCA for the analysis of multicate-gory data was reported by Page [10]. This involved the characterization of orange juice samples of different varieties from various geographical locations using the following techniques: HPLC (34), inductively coupled plasma emission spectrometry (10), infrared (40), fluorescence (6), ^{13}C NMR (49), ultraviolet (29), enzymatic analysis (6), and GC (13). The numbers in brackets give the number of parameters measured using each method; the use of such a variety of analytical techniques allowed the identification of different components of the juices, e.g. trace elements, sugars, organic acids, polyphenols etc. The performance of SIMCA on the HPLC data (carotenoids, flavones, and flavonoids) is shown in Table 7.5. This is a particular type of table that is often used to

Table 7.5 SIMCA classification of orange juice samples (reproduced from ref. [10] with permission from the Institute of Food Technologists).

True class[a]	Computed class – assigned (%)							
	1	2	3	4	5	6	7	8
1	100							
2		100						
3			82.4			11.8		5.9
4				94.4	5.6			
5				6.7	93.3			
6				16.7		83.3		
7							100	
8				18.2				81.8

[a]The classes are: 1, Valencia (Florida); 2, Valencia (other US); 3, Valencia (non-US); 4, Hamlin (all); 5, Pineapple (all); 6, Navel (all); 7, others (Florida); 8, others (Brazil).

summarize the results of classification algorithms and which has the rather wonderful name 'confusion matrix' as discussed in the next section. In this example SIMCA can be seen to have done quite an impressive job in classification; three of the classes (1, 2 & 7) are completely correctly predicted and two others (4 & 5) are over 90 % correct. The lowest figure on the diagonal of the table is 81.8 %, and in most cases, wrong assignment is only made to one other class.

7.2.3   Confusion Matrices

Reporting the results of a classification method might, at first thought, appear to be simple. All that is required is to report the number, or percentage, of each class that is correctly predicted, surely? Unfortunately, on reflection, it isn't quite as simple as that. Consider the situation where we have an equal number of samples of two classes and that the classification method, discriminant analysis, k-nearest-neighbour, SIMCA or whatever, correctly classifies all of the members of class one but gets all the members of class two wrong. The prediction success for class one is 100 % but the overall prediction rate is only 50 %. This is where the confusion matrix comes in. Table 7.6 shows the layout of a confusion matrix.

In this table the two classes are conventionally referred to as positive and negative although for this one can read active/inactive or class 1/class two or big/small and so on. The first entry in the table, TP, is the number of true positives, that is to say the number of members of the positive

Table 7.6 Confusion matrix.

True class	Predicted			
positive	TP	FN	TP+FN	100*TP/(TP+FN)
negative	FP	TN	FP+TN	100*TN/(TN+FP)
	TP+FP	FN+TN	TP+TN+FP+FN	100*(TP+TN)/(TP+TN+FP+FN)
	100*TP/(TP+FP)	100*TN/(FN+TN)		

class which are correctly classified as positive. The next entry on this row, FN, stands for false negatives, in other words the number of members of the positive class which are predicted to be negative. The second row contains the corresponding false positives, FP, and true negatives, TN, and the third row contains totals of positive and negative predictions. The top two entries of the last column are known as *sensitivity*, the percentage of positives predicted correctly, and *specificity*, the percentage of negatives predicted correctly. These are the numbers that are most often considered or quoted when assessing or reporting the results of a classification analysis. The entry in the bottom right hand cell of the table is the total percentage correctly predicted and this is also often used to summarize the results of an analysis. The bottom row contains two important numbers and these are the percentage of cases correctly predicted to be positive, known as the *predictive power of a positive test*, and the corresponding percentage of cases correctly predicted to be negative, the *predictive power of a negative test*.

When considering how well a classification method has performed on a particular data set it is important to be aware of these predictive powers as well as the sensitivity and specificity of predictions. There may be circumstances where false positives or negatives are less important, for example it may be better to mis-classify a blood sample as indicative of cancer since further tests will identify this, but generally one might aim to have predictive powers and sensitivity/specificity of about the same order. It is also important to be aware of class membership numbers since this affects the expected ability of a classifier, as discussed before for k-nearest neighbour methods, and also the number of descriptors used in a classification model (see next section). Some statistical packages may produce a simplified confusion matrix as shown in Figure 7.5. The results shown in the figure, termed a classification matrix, contain just the totals for each class and the percentages correct without the predictive powers of the positive and negative tests. The figure also shows results for what is termed a jacknife classification matrix. Jacknife is another name for leave-one-out cross-validation (see Section 6.4.1) in which each sample is left out in turn, the discriminant function fitted to the remaining samples and the left out sample classified. In this case the results are identical indicating that the function is able to generalize, or that none of the samples is an outlier, or both, of course. As discussed in the last chapter this is not necessarily the best test of predictive ability of any mathematical model and it is much better to split the data into training and test sets, preferably several times, as a check.

```
Classification matrix (cases in row categories classified into columns)
                    1           2           3   %correct
1                  50           0           0       100
2                   0          48           2        96
3                   0           1          49        98
    Total          50          49          51        98


Jackknifed classification matrix
                    1           2           3   %correct
1                  50           0           0       100
2                   0          48           2        96
3                   0           1          49        98
    Total          50          49          51        98
```

Figure 7.5 Example of the output from discriminant analysis applied to a well known data set (the Fisher Iris data).

7.2.4 Conditions and Cautions for Discriminant Analysis

As was said at the beginning of this chapter, supervised learning techniques in general are subject to the dangers of chance effects, and discriminant techniques are no exception. Jurs and co-workers have reported quite extensive investigations of the problem of chance separation [11–14]. As was the case for regression analysis using random numbers, it was found that the probability of achieving a separation by chance using linear discriminant analysis increased as the number of variables examined was increased. The situation, however, for discriminant analysis is compounded by the question of the dimensionality of the data set. In the limit where a data set contains as many physicochemical (or structural, e.g. indicators) variables as there are samples in the set, there is a trivial solution to the discriminant problem. The discriminant procedure has as many adjustable parameters (the discriminant function coefficients) as there are data points and thus can achieve a perfect fit. This is equivalent to fitting a multiple linear regression model to a data set with zero degrees of freedom. The recommendation from Jurs's work is that the ratio of data points (compounds, samples, etc.) to descriptor variables should be three or greater.

Another aspect of the dimensionality of a data set that is perhaps not quite so obvious concerns the number of members in each class. Ideally, each of the classes in a data set should contain about the same number of members. If one class contains only a small number of samples, say 10 % of the total points or less, then the discriminant function may be able to achieve a trivial separation despite the fact that the ratio of

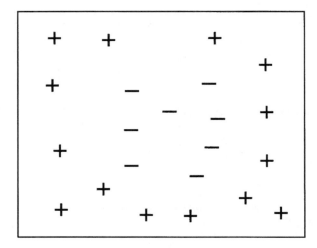

Figure 7.6   Illustration of a data set in which one class is embedded in another.

points to descriptors for the overall data set is greater than three. The following guidelines should be borne in mind when applying discriminant techniques.

- The number of variables employed should be kept to a minimum (by preselection) and the ratio $N:P$ (samples : parameters) should be greater than three.
- The number of members in each class should be about equal, if necessary by changing the classification scheme or by selecting samples.

Finally, it may be the case that the data is not capable of linear separation. Such a situation is shown in Figure 7.6 where one class is embedded within the other. An interesting technique for the treatment of such data sets makes use of PCs scaled according to the parameter values of the class of most interest, usually the 'actives' [15]. This is somewhat reminiscent of the SIMCA method.

7.3   REGRESSION ON PRINCIPAL COMPONENTS AND PARTIAL LEAST SQUARES

Methods such as PCA (see Section 4.2) and factor analysis (FA) (see Section 5.3) are data-reduction techniques which result in the creation of new variables from linear combinations of the original variables. These new variables have an important quality, orthogonality, which makes them particularly suitable for use in the construction of regression models. They are also sorted in order of importance, in so far as the amount of variance in the independent variable set that they explain, which makes the choice of them for regression equations somewhat easier than the choice of 'regular' variables. Unfortunately, the fact that they are linear combinations makes the subsequent interpretation of regression models somewhat more difficult. The following sections describe regression using PCs, a variant of this called partial least squares (PLS), and, briefly, a technique called continuum regression which embraces both of these and ordinary multiple regression.

7.3.1   Regression on Principal Components

The first step in carrying out principal component regression (PCR) is, unsurprisingly, a PCA. This produces scores and loadings as described by Equation (4.1), reproduced below.

$$PC_1 = a_{1,1}v_1 + a_{1,2}v_2 + \ldots\ldots a_{1,P}v_P$$
$$PC_2 = a_{2,1}v_1 + a_{2,2}v_2 + \ldots\ldots a_{2,P}v_P$$
$$PC_q = a_{q,1}v_1 + a_{q,2}v_2 + \ldots\ldots a_{q,P}v_P \qquad (7.4)$$

The scores are the values of each PC for each sample, the loadings are the subscripted coefficients $(a_{i,j})$ in Equation (7.4). A score for a particular compound or sample in the data set (for a particular principal component) is computed by multiplying the descriptor variable values by the appropriate loadings and then adding together the products. Each PC has associated with it a quantity called an eigenvalue, a measure of how much of the variance in the original data set is described by that component. Since the components are calculated in order of decreasing amounts of variance explained, it follows that the first PC will have the largest eigenvalue in the set and subsequent PCs successively smaller eigenvalues. Can these eigenvalues be used as a measure of importance or 'significance'? After all, PCA will produce as many components as the smaller of N points or P dimensions,[4] so there may be as many PCs as there were originally dimensions. The answer to this question is a reassuring perhaps! If the original data are autoscaled (see Section 3.3),

---

[4] Actually, it is the rank of the matrix, denoted by $r(\mathbf{A})$, which is the maximum number of linearly independent rows (or columns) in $\mathbf{A}$. $0 \leq r(\mathbf{A}) \leq \min(n,p)$, where $\mathbf{A}$ has n rows and p columns.

then each variable will contribute one unit of variance to the data set, which will have a total variance of P where P is the number of variables in the set. As PCs are produced, their eigenvalues will decrease until they fall below a value of one. At this point the components will no longer be explaining as much variance as one of the original variables in the set and this might make a reasonable limit to assess components as meaningful (however, see later). For most real data sets, components with an eigenvalue of approximately one are found in a far smaller number than the original number of properties.

Having carried out PCA, what comes next? The principal component scores are treated as any other variables would be in a multiple regression analysis and MLR models are constructed as shown in Equation (7.5)

$$y = a_1 P C_{x1} + a_2 P C_{x2} + \ldots \ldots a_p P C_{xp} + c \qquad (7.5)$$

where y is some dependent variable, perhaps $\log 1/C$ for a set of biological results, a_1 to a_p are a set of regression coefficients fitted by least squares to the p principal components in the model and c is a constant (intercept). The fit of the regression model can be evaluated by using the usual regression statistics and the equation can be built up by forward-inclusion, backward-elimination or whatever (see Chapter 6). Is it possible to say which PCs should be incorporated into a PCR model? Surely the components are calculated in order of their importance and thus we might expect them to enter in the order one, two, three, and so on. This is partly true, components are calculated in order of their importance in terms of explaining the variance in the set of *independent* variables and very often the first one or two components will also be best correlated with a dependent variable. But for this to happen depends on a good choice of variables in the first place, in so far as they are correlated with y, and the fact that a linear combination of them will correlate with the dependent variable.

An example may illustrate this. In an attempt to describe the experimentally determined formation constants for charge-transfer complexes of monosubstituted benzenes with trinitrobenzene, a set of computational chemistry parameters were calculated [16]. The initial data set contained 58 computed physicochemical descriptors, which after the removal of correlated variables (see Section 3.5), left a set of 31. Parameters were selected from this set, on the basis of their ability to describe the formation constants (see Section 7.4), to leave a reduced subset of 11 descriptors. PCA carried out on this set gave rise to four components with

Table 7.7 Variable loadings* for the first five PCs derived from the reduced data set of 11 variables (reproduced from ref. [16] with permission of the Royal Society of Chemistry).

	Component (eigenvalue)				
	1 (2.73)	2 (2.19)	3 (1.78)	4 (1.23)	5 (0.95)
Variable			Loading		
CMR	0.48	−0.34			
$c$logP		−0.41		−0.47	
E_{HOMO}		−0.36		0.49	
P3	0.41				0.33
μ_x		0.48		−0.37	
$Sn(1)$	−0.31				0.42
$Sn(2)$			−0.59		
P1				−0.41	0.60
$Fe(4)$	−0.39			−0.38	
μ	−0.40	0.40			0.38
$Sn(3)$			−0.60		

*Only loadings above 0.3 are shown for clarity.

an eigenvalue greater than one, and one component with an eigenvalue close to one (0.95), as shown in Table 7.7.

Forward-inclusion regression analysis between κ, a substituent constant derived from the formation constants, and these PCs led to the following equations

$$\kappa = 0.191 PC1 + 0.453 \tag{7.6}$$
$$R^2 = 0.5 \quad F = 33.01 \quad SE = 0.32$$

$$\kappa = 0.191 PC1 + 0.193 PC4 + 0.453 \tag{7.7}$$
$$R^2 = 0.732 \quad F = 43.77 \quad SE = 0.24$$

$$\kappa = 0.191 PC1 + 0.193 PC4 + 0.130 PC5 + 0.453 \tag{7.8}$$
$$R^2 = 0.814 \quad F = 45.22 \quad SE = 0.20$$

$n = 35$ for all three equations, the F statistics are all significant (at 99 % probability) and the t statistics for the individual regression coefficients are significant at greater than the 1 % level.

A number of things may be seen from these equations. The first component to be incorporated in the regression models was indeed the first PC and this, combined with a constant, accounted for half of the variance in the dependent variable set ($R^2 = 0.5$). The next component to enter the model, however, was PC_4 despite the fact that the variables in the set of 11 had been chosen for their *individual* ability to describe κ. Clearly, the linear combinations imposed by PCA, combined with the requirements of orthogonality, did not produce new variables in PC_2 and PC_3 which were useful in the description of κ. The third PC to be included has an eigenvalue of less than one and yet it is seen to be significant in Equation (7.8). If the eigenvalue cut-off of less than one had been imposed on this data set, Equation (7.8) would not have been found.

As extra terms are added to the regression model it can be seen that the regression coefficients do not change, unlike the case for MLR with untransformed variables where collinearity and multicollinearity amongst the descriptors can lead to instability in the regression coefficients. The regression coefficients in Equations (7.6) to (7.8) remain constant because the principal component scores are orthogonal to one another, the inclusion of extra terms in a PCR leads to the explanation of variance in the dependent variable not covered by terms already in the model. These features of PCR make it an attractive technique for the analysis of data; an unsupervised learning method (lower probability of chance effects?) produces a reduced set of well-behaved (orthogonal) descriptors followed by the production of a stable, easily interpreted, regression model. Unfortunately, although the regression equation is easily understood and applied for prediction, the relationship with the original variables is much more obscure. This is the big disadvantage of PCR, and related techniques such as PLS described in the next section. Inspection of Table 7.7 allows one to begin to ascribe some chemical 'meaning' to the PCs, for example, bulk factors (CMR and P3) load onto PC_1 and log P (-0.47) loads onto PC_4, but it should be remembered that the PCs are mathematical constructs designed to explain the variance in the x set. The use of varimax rotation (see Section 4.2) may lead to some simplification in the interpretation of PCs.

7.3.2 Partial Least Squares

The method of partial least squares (PLS) is also a regression technique which makes use of quantities like PCs derived from the set of independent variables. The PCs in PLS regression models are called latent variables (LV),[5] as shown in the PLS equation, Equation (7.9).

$$y = a_1 LV_1 + a_2 LV_2 + \ldots\ldots a_p LV_p \tag{7.9}$$

---

[5] This term may be generally used to describe variables derived from measured variables (for example, PCs, factors, etc.).

where y is a dependent variable and a_1 to a_p are regression coefficients fitted by the PLS procedure. Each latent variable is a linear combination of the independent variable set.

$$LV_1 = b_{1,1}x_1 + b_{1,2}x_2 + \ldots \ldots b_{1,p}x_p$$
$$LV_2 = b_{2,1}x_1 + b_{2,2}x_2 + \ldots \ldots b_{2,p}x_p$$
$$\ldots\ldots\ldots\ldots\ldots\ldots\ldots\ldots\ldots\ldots\ldots\ldots$$
$$LV_q = b_{q,1}x_1 + b_{q,2}x_2 + \ldots \ldots b_{q,p}x_p \qquad (7.10)$$

As in PCA, PLS will generate as many latent variables (q) as the smaller of P (dimensions) or N (samples). Thus far, PLS appears to generate identical models to PCR so what is the difference (other than terminology)? The answer is that the PLS procedure calculates the latent variables and the regression coefficients in Equation (7.9) all at the same time. The algorithm is actually an iterative procedure [17] but the effect is to combine the PCA step of PCR with the regression step. Latent variables, like PCs, are calculated to explain most of the variance in the x set while remaining orthogonal to one another. Thus, the first latent variable (LV_1) will explain most of the variance in the independent set, LV_2 the next largest amount of variance and so on. The important difference between PLS and PCR is that the latent variables are constructed so as to maximize their covariance with the dependent variable. Unlike PCR equations where the PCs do not enter in any particular order (see Equations (7.6) to (7.8)) the latent variables will enter PLS equations in the order one, two, three, etc. The properties of latent variables are summarized below.

- The first latent variable explains maximum variance in the independent set; successive latent variables explain successively smaller amounts of variance.
- The latent variables conform to 1 with the provision that they maximize their covariance with the dependent variable.
- The latent variables are orthogonal to one another.

One problem with the PLS procedure, common to both PCR and multiple linear regression (MLR), is the choice of the number of latent variables to include in the model. The statistics of the fit can be used to judge the number of variables to include in an MLR but the situation is somewhat more complex for PCR and PLS. Judgement has to be exercised as to how 'significant' the LVs or PCs are. Although the statistics of the fit may indicate that a particular PC or LV is making a significant contribution to a regression equation, that variable may contain very little 'information'. The eigenvalue of a PC or LV may be a guide but as was seen in the previous section, some cut-off value for the eigenvalue is not necessarily a good measure of significance.

A commonly used procedure in the construction of PLS models is to use leave-one-out (LOO) cross-validation (see Section 6.4.1) to estimate prediction errors. This works by fitting a PLS model to $n - 1$ samples and making a prediction of the y value for the omitted sample ($\hat{y}$). When this has been carried out for every sample in the data set a predicted residual error sum of squares (PRESS) can be calculated for that model.

$$\text{PRESS} = \sum_{i=1}^{n} (y_i - \hat{y}_i)^2 \qquad (7.11)$$

Note that this sum of squares looks similar to the residual sum of squares (RSS) given by Equation (6.12) but is different; in Equation (6.12) the $\hat{y}_i$ is predicted from an equation that includes that data point; here the $\hat{y}_i$ is not in the model hence the term *predictive* residual sum of squares. PRESS values are calculated for all the models to be considered and then various criteria may be employed to determine the optimal model. One simple way to do this is to choose the model with the lowest PRESS value. Since PRESS is a function of residuals this model will minimize predictive errors. Another way is to select a model which yields a local minimum. The model is chosen to contain the fewest components while minimizing PRESS. A plot of PRESS versus the number of components (a scree plot) is quite often illuminating in these situations, as PRESS will decrease with increasing components and then will begin to increase, as predictive ability worsens, but may then decrease again. A third method is to set some threshold value of PRESS and the optimal model is then chosen to be the first model with a PRESS score below this threshold. These criteria, particularly the threshold value, however are somewhat subjective.

Various numeric criteria involving PRESS have also been proposed [18]. Wold [17] suggested a quantity called the E statistic (Equation (7.12)) to judge the difference in predictive ability of two PLS models.

$$E = \frac{\text{PRESS}_i}{\text{PRESS}_{i-1}} \qquad (7.12)$$

The E statistic compares a PLS model of i components with the model containing one component less and in order to evaluate E for the one component model, PRESS for the model containing no components is calculated by comparing predicted values with the mean. The original suggestion for using E was that models with an E value < 1.0 were significant. Some problems with leave-one-out methods have already been discussed (Section 6.4.1) and thus other approaches to model selection have involved different choices of selection for the left out set [19]. All of this may appear very confusing but fortunately most implementations of PLS offer some built-in model selection criteria and the usual principle of parsimony, that is to say selection of the simplest model where possible, is often the best advice.

On the face of it, PLS appears to offer a much superior approach to the construction of linear regression models than MLR or PCR (since the dependent variable is used to construct the latent variables) and for some data sets this is certainly true. Application of PLS to the charge-transfer data set described in the last section resulted in a PLS model containing only two dimensions which explained over 90 % of the variance in the substituent constant data. This compares very favourably with the two- and three-dimensional PCR equations (Equations (7.7) and (7.8)) which explain 73 and 81 % of the variance respectively. Another advantage that is claimed for the PLS approach is its ability to handle redundant information in the independent variables. Since the latent variables are constructed so as to correlate with the dependent variable, redundancy in the form of collinearity and multicollinearity in the descriptor set should not interfere. This is demonstrated by fitting PLS models to the 31 variable and 11 variable parameter sets for the charge-transfer data. As shown in Table 7.8 the resulting PLS models account for very similar amounts of variance in κ.

How are PLS models used? One obvious way is to simply make predictions for test set samples by calculation of their latent variables from Equation (7.10) and application of the appropriate regression coefficients (Equation (7.9)). The latent variables may be used like PCs for data display (see Chapter 4) by the construction of scores plots for samples and

Table 7.8 Modelling κ by PLS (reproduced from ref. [16] with permission of the Royal Society of Chemistry).

PLS model of dimension:	Percentage of κ variance explained using:	
	11 variable dataset	31 variable dataset
1	78.6	78.7
2	92.9	90.4
3	94.9	95.1

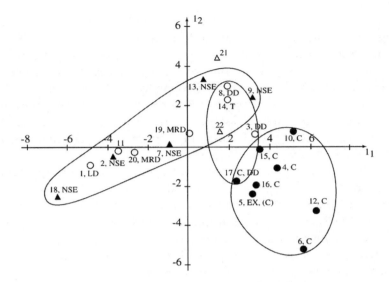

Figure 7.7 A PLS scores plot for a set of halogenated ether anaesthetics; the ellipses enclose compounds with similar side-effects (reproduced from ref. [20] with permission of Wiley-VCH).

loadings plots for variables. A PLS analysis of halogenated ether anaesthetics allowed the production of the scores plot shown in Figure 7.7 in which anaesthetics with similar side-effects are grouped together [20]. The biological data available for these compounds included a measure of toxicity and separate PLS models were fitted to the anaesthetic and toxicity data. Prediction of toxicity was good from a three-component PLS model as shown in Figure 7.8.

Another widespread use of the PLS technique is based on its ability to handle very large numbers of physicochemical parameters. The increasing use of molecular modelling packages in the analysis of biological and other data has led to the establishment of so-called three-dimensional QSAR [21–23]. In these approaches a grid of points is superimposed on each of the molecules in the training set. Probes are positioned at each of the points on the grid and an interaction energy calculated between the probe and the molecule. Depending on the resolution chosen for the grid, several thousand energies may be calculated for each type of probe for every molecule in the set. Clearly, many of these energies will be zero or very small and may be discarded, and many will be highly correlated with one another. For example, when a positively charged probe is placed at the grid points near a region of high electron density, the attractive interactions will be similar. PLS is used to model the relationship between these grid point interaction energies and the dependent

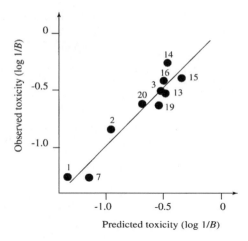

Figure 7.8   Plot of toxicity predictions from a three-component PLS model (reproduced from ref. [20] with permission of Wiley-VCH).

variable. The resultant PLS models may be visualized by displaying the 'important' grid points, as determined by their loadings onto the latent variables and the coefficients of these variables in the PLS regression equation.

What are the problems with the PLS technique? One problem, like the question of deciding dimensionality, is shared with MLR and PCR. Why fit a linear model? The imposition of simple linear relationships on nature might be thought of as 'expecting too much'. Fortunately, simple or at least fairly simple linear relationships often do hold and linear models do quite a good job. Another problem with PLS is also shared with PCR and that is the question of interpretation of the latent variables. Table 7.9 shows the important loadings of the 11 variable descriptor set for the charge-transfer data onto the first two PCs and the first two LVs. LV_1 is very similar to PC_1 with the exception that it contains $clogP$, LV_2 has some similarity with PC_2 but of course PC_2 was not included in the PCRs.

7.3.3   Continuum Regression

In MLR the equations are constructed so as to maximize the explanation of the correlation between the dependent variable and the independent variables. Variance in the independent set is ignored, regression coefficients are simply calculated on the basis of the fit of y to the x variables.

Table 7.9 PC and LV* loadings for charge-transfer data (reproduced from ref. [16] with permission of the Royal Society of Chemistry).

Variable	PCR component (eigenvalue)		PLS latent variable	
	1 (2.73)	2 (2.19)	1	2
	Loading		Loadings	
CMR	0.48	−0.34	0.48	
$clogP$		−0.41	−0.32	0.67
E_{HOMO}		−0.36		−0.51
P3	0.41		0.42	
μ_x		0.48		0.36
$S_n(1)$	−0.31		−0.24	
$S_n(2)$				
P1				
$F_e(4)$	−0.39		−0.34	
μ	−0.40	0.40	−0.39	0.4
$S_n(3)$				

*As in Table 7.7, only loadings above 0.3 are shown for clarity (except the loading for $S_n(1)$ on latent variable 1 so that it can be compared with PC_1).

PCR, on the other hand, concentrates on the explanation of variance in the descriptor set. The first step in PCR is the generation of the PCs, regression coefficients are calculated on the basis of explanation of the correlation between y and these components.

These two approaches to the construction of regression models between y and an x set can be viewed as extremes. The relative balance between explanation of variance (in the x set) and correlation (y with x) can be expressed as a parameter, α, which takes the value of 0 for MLR and 1 for PCR. PLS regression sets out to describe both variance and correlation, and thus will have a value of α of 0.5, midway between these two extremes. Continuum regression (CR) is a new type of regression procedure which contains an adjustable parameter, α, which allows the production of all three types of regression model [24]. An alternative formulation of continuum regression has been developed in which the parameter α is optimized during the production of the regression model [18]. The two most popular forms of regression analysis, MLR and PLS, tend to be applied to data sets in a quite arbitrary way, often dictated by the whim (or experience) of the analyst. Continuum regression presents the opportunity to allow the structure within a data set to determine the most appropriate value of α, and hence the decision as to which regression model to fit. Indeed, since α is an adjustable parameter

Table 7.10 Continuum regression results for literature data.

Reference[a]	Number of samples	Number of variables	R^2	α[b]
[25] (PLS)	7	11	0.69 (1)[c]	0.73
[26] (MLR)	21	6	0.94 (3)	0.35, 0.07, 0.07
[27] (MLR)	25	3	0.85 (1)	0.55
[28] (MLR)	40	4	0.95 (3)	0.58, 0.85, 0.58
[29] (MLR)	12	3	0.85 (2)	0.4, 0.24

[a]The method used in the original report is shown in brackets.
[b]Where more than one component is used, the α values are for models of each dimensionality.
[c]The number of components used for this R^2 is given in brackets.

which can adopt any value between zero and one, it is possible to fit models which do not correspond to MLR, PLS, or PCR. This is illustrated in Table 7.10 for some literature data sets where it can be seen that α values of 0.24, 0.35, 0.73, and 0.85 are obtained for some components of the fitted models. The first two of these correspond to models which are somewhere between MLR and PLS, while the latter two correspond to models in between PLS and PCR. The two- and three-dimensional models for the Wilson and Famini example [26] have α values near zero which correspond to the MLR used in the original analysis. The example data set from Clark and co-workers [27], on the other hand, which was also originally fitted by MLR gives an α value of 0.55 corresponding to a PLS model.

The charge-transfer data shown in the last two sections has also been modelled by continuum regression. Since it was possible to fit both PCR and PLS models to this data set it was expected that CR should model this data well, as indeed it does. Table 7.11 shows a summary of these models along with the corresponding information from the PCR and PLS models.

It can be seen from the table that continuum regression is doing an even better job at modelling the data set than PLS which, itself, was an improvement over PCR. The correlation coefficient for the first dimensional models was 0.5, 0.786 and 0.84 for PCR, PLS and CR respectively. PCR had to be taken to 3 dimensions to get a reasonable model whereas both PLS and CR modelled the set well in just two dimensions although it should be remembered that each of these dimensions contains loadings

Table 7.11 Summary of PCR, PLS and continuum regression models.

	PCR			PLS			Continuum regression		
Model Dimension	1	2	3	1	2	3	1	2	3
Components	PC1	PC1	PC1	LV1	LV1	LV1	C1	C1	C1
		PC4	PC4		LV2	LV2		C2	C2
			PC5			LV3			C4
R^2	0.5	0.732	0.814	0.786	0.929	0.949	0.84	0.95	0.97
α	1	1	1	0.5	0.5	0.5	0.31	0.31	0.31
								0.24	0.24
									0.46

for all of the original 11 variables. The α values for the PCR and PLS models are fixed of course, at 1.0 and 0.5 respectively, so it is interesting to see the values of α assigned by the algorithm to the three CR components. The first two represent something in between MLR ($\alpha = 0$) and PLS, while the third component is much closer to a PLS latent variable. Interpretation of these α values is not obvious but it is intriguing that continuum regression appears to offer the useful facility of allowing the data to select the model to fit. Finally, the (simplified) structure of the first two CR components is shown in Table 7.12 for comparison with the PCR and PLS components shown in Table 7.9.

7.4 FEATURE SELECTION

One of the problems involved in the analysis of any data set is the identification of important features. So far this book has been involved with the independent variables, but the problem of feature selection may also apply to a set of dependent variables (see Chapter 8). The identification of redundancy amongst variables has already been described (Section 3.5) and techniques have been discussed for the reduction of dimensionality as a step in data analysis. These procedures are unsupervised learning methods and thus do not use the property of most interest, the dependent variable. What about supervised learning methods? The obvious way in which a supervised technique may be used for the identification of important features is to examine the relationship between the dependent and independent variables. If the dependent variable is

Table 7.12 Constitution of CR components.

Variable	CR component 1	CR component 2
	Loading*	
CMR	0.47	
$c\log P$	−0.40	−0.80
E_{HOMO}		−0.39
P3	0.40	
μ_x	0.37	
$S_n(1)$		
$S_n(2)$		
P1		
$F_e(4)$	−0.3	
μ	0.30	
$S_n(3)$		

* As in Table 7.7, only loadings above 0.3 are shown for clarity.

continuous then correlation coefficients may be calculated, if classified then variables can be selected which split the data (more or less) into the two or more classes. Both of these methods have already been mentioned; the 11 parameters from the charge-transfer set were selected by their individual correlation with κ and the gas chromatography parameters from the fire ants were chosen for their ability to distinguish the two species.

This approach selects variables based on their individual usefulness but of course they are often then combined into overall predictive models. Thus, another supervised means of selecting variables is to examine the parameters that are included in the models. Significant terms in MLR equations may point to important variables as may high loadings in PCR components or PLS latent variables. High loadings for variables in the latter are likely to be more reliable indicators of importance since the PLS variables are constructed to be highly correlated with the dependent variable. This process may be useful if further modelling, say the use of non-linear methods such as artificial neural networks, is going to be applied to the data set. LDA models may also be used to identify important variables but here it should be remembered that discriminant functions are not unique solutions. Thus, the use of LDA for variable selection may be misleading.

Genetic algorithms have already been mentioned (Section 6.3.1.5) as a means of building multiple linear regression models from large pools of potential descriptor variables and this process, in itself, may be used as a means of variable selection. Since the genetic approach often gives rise to a population of models then the frequency of occurrence of variables in the population may be used as a measure of variable importance. Genetic methods may be used as a 'wrapper' around many sorts of models, such as linear discriminant functions, PLS equations, artificial neural networks and so on, and thus this technique can be generally employed for variable selection. Whatever form of supervised learning method is used for the identification of important variables it is essential to bear in mind one particular problem with supervised learning: chance effects. In order to reduce the probability of being misled by chance correlations it is wise to be conservative in the use of supervised learning techniques for variable selection.

7.5 SUMMARY

This chapter has described some of the more commonly used supervised learning methods for the analysis of data; discriminant analysis and its relatives for classified dependent data, variants of regression analysis for continuous dependent data. Supervised methods have the advantage that they produce predictions, but they have the disadvantage that they can suffer from chance effects. Careful selection of variables and test/training sets, the use of more than one technique where possible, and the application of common sense will all help to ensure that the results obtained from supervised learning are useful.

In this chapter the following points were covered:

1. the modelling and prediction of classified data using discriminant analysis;
2. modelling classified data with SIMCA;
3. reporting results with a confusion matrix;
4. regression on principal components (PCR);
5. partial least squares regression (PLS);
6. model selection for PCR and PLS;
7. continuum regression;
8. feature selection using supervised methods.

REFERENCES

[1] Dillon, W.R. and Goldstein, M. (1984). *Multivariate Analysis Methods and Applications*, pp. 360–93. John Wiley & Sons, Inc., New York.

[2] Martin, Y.C., Holland, J.B., Jarboe, C.H., and Plotnikoff N. (1974). *Journal of Medicinal Chemistry*, **17**, 409–13.

[3] Brill, J.H., Mayfield, H.T., Mar, T., and Bertsch, W. (1985). *Journal of Chromatography*, **349**, 31–8.

[4] Zalewski, R.I. (1992). *Journal de Chimie Physique et de Physico-Chemie Biologique*, **89**, 1507–16.

[5] Kier, L.B. (1980). *Journal of Pharmaceutical Science*, **69**, 416–19.

[6] Lanteri, S., Conti, P., Berbellini, A., *et al.* (1988). *Journal of Chemometrics*, **3**, 293–9.

[7] Wold, S. and Sjostrom, M. (1977). In *Chemometrics – Theory and Application*, B.R. Kowalski (ed.), p. 243. American Chemical Society, Washington D.C.

[8] Dunn, W.J., Wold, S., and Martin, Y.C. (1978). *Journal of Medicinal Chemistry*, **21**, 922–30.

[9] Scarmino, I.S., Bruns, R.E., and Zagatto, E.A.G. (1982). *Energia Nuclear e Agricultura*, **4**, 99–111.

[10] Page, S.W. (1986). *Food Technology*, **Nov.**, 104–9.

[11] Stuper, A.J. and Jurs, P.C. (1976). *Journal of Chemical Information and Computer Sciences*, **16**, 238–41.

[12] Whalen-Pedersen, E.K. and Jurs, P.C. (1979). *Journal of Chemical Information and Computer Sciences*, **19**, 264–6.

[13] Stouch, T.R. and Jurs, P.C. (1985). *Journal of Chemical Information and Computer Sciences*, **25**, 45–50.

[14] Stouch, T.R. and Jurs, P.C. (1985). *Journal of Chemical Information and Computer Sciences*, **25**, 92–8.

[15] Rose, V.S., Wood, J., and McFie, H.J.H. (1991). *Quantitative Structure–Activity Relationships*, **10**, 359–68.

[16] Livingstone, D.J., Evans, D.A., and Saunders, M.R. (1992). *Journal of the Chemical Society-Perkin Transactions II*, 1545–50.

[17] Wold, S. (1978). *Technometrics*, **20**, 397–405.

[18] Malpass, J.A., Salt, D.W., Ford, M.G., Wynn, E.W. and Livingstone D.J. (1994). In *Advanced Computer-assisted Techniques in Drug Discovery*, H. Van de Waterbeemd (ed.), Vol 3 of *Methods and Principles in Medicinal Chemistry*, R. Mannhold, P. Krogsgaard-Larsen and H.Timmerman (eds), pp. 163–89, VCH, Weinheim.

[19] Livingstone, D.J. and Salt, D.W. (2005). Variable selection – spoilt for choice? in *Reviews in Computational Chemistry*, Vol 21, K. Lipkowitz, R. Larter and T.R. Cundari (eds), pp. 287–348, Wiley-VCH.

[20] Hellberg, S., Wold, S., Dunn, W.J., Gasteiger, J., and Hutchings, M.G. (1985). *Quantitative Structure–Activity Relationships*, **4**, 1–11.

[21] Goodford, P.J. (1985). *Journal of Medicinal Chemistry*, **28**, 849–57.

[22] Cramer, R.D., Patterson, D.E., and Bunce, J.D. (1988). *Journal of the American Chemical Society*, **110**, 5959–67.

[23] Kubinyi, H. (ed.) (1993). *3D QSAR in Drug Design: Theory, Methods, and Applications*. ESCOM, Leiden.

[24] Stone, M. and Brooks, R.J. (1990). *Journal of the Royal Statistical Society Series B-Methodological*, **52**, 237–69.

[25] Dunn, W.J., Wold, S., Edlund, U., Hellberg, S., and Gasteiger, J. (1984). *Quantitative Structure–Activity Relationships*, **3**, 134–7.

[26] Wilson, L.Y. and Famini, G.R. (1991). *Journal of Medicinal Chemistry*, **34**, 1668–74.

[27] Clark, M.T., Coburn, R.A., Evans, R.T., and Genco, R.J. (1986). *Journal of Medicinal Chemistry*, **29**, 25–9.

[28] Li, R., Hansch, C., Matthews, D., Blaney, J.M., *et al.* (1982). *Quantitative Structure–Activity Relationships*, **1**, 1–7.

[29] Diana, G.D., Oglesby, R.C., Akullian, V., *et al.* (1987). *Journal of Medicinal Chemistry*, **30**, 383–8.

8

Multivariate Dependent Data

Points covered in this chapter

- Use of multiple dependent data in PCA and factor analysis
- Cluster analysis and multiple dependents
- Biplots
- Spectral map analysis
- Methods to handle multiple responses and multiple independents

8.1 INTRODUCTION

The last four chapters of this book have all been concerned with methods that handle multiple independent (descriptor) variables. This has included techniques for displaying multivariate data in lower dimensional space, determining relationships between points in P dimensions and fitting models between multiple descriptors and a single response variable, continuous or discrete. Hopefully, these examples have shown the power of multivariate techniques in data analysis and have demonstrated that the information contained in a data set will often be revealed only by consideration of all of the data at once. What is true for the analysis of multiple descriptor variables is also true for the analysis of multiple response data. All of the techniques so far described for independent variables may also be applied to multiple dependent variables, as illustrated in Figure 8.1.

The following sections of this chapter demonstrate the use of principal components and factor analysis (FA) in the treatment of multiple

A Practical Guide to Scientific Data Analysis   David Livingstone
© 2009 John Wiley & Sons, Ltd

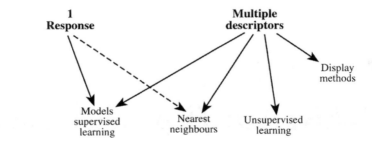

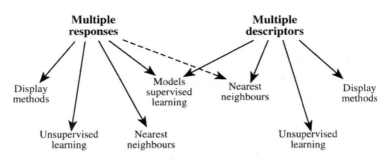

Figure 8.1 Analytical methods that may be applied to one response with multiple descriptors and multiple responses with multiple descriptors.

response data, cluster analysis, and a method called spectral map analysis. The last section discusses the construction of models between multiple dependent and independent variable sets. It is perhaps worth pointing out here that the analysis of multivariate response data is even more unusual in the scientific literature than the multivariate analysis of independent data sets. This is probably not only a reflection of the unfamiliarity of many multivariate techniques but also of the way in which experiments are conducted. It is not uncommon to design experiments to have only one outcome variable, an easily determined quantity such as colour, taste, percentage inhibition, and so on. This demonstrates a natural human tendency to try to make a complicated problem (the world) less complicated. In many cases, our experiments do generate multiple responses but these are often discarded, or processed so as to produce a single 'number', because of ignorance of methods which can be used to handle such data. Perhaps the following examples will show how such data sets may be usefully treated.

8.2 PRINCIPAL COMPONENTS AND FACTOR ANALYSIS

Compounds which are biologically active often show different effects in different tests, e.g. related receptor assays, as shown later, or in different species. A very obvious example of this is the activity of antibacterial compounds towards different types of bacteria. If a biological results table is drawn up in which each row represents a compound and each column the effect of that compound on a different bacterial strain, principal components analysis (PCA) may be used to examine the data. An example of part of such a data set, for a set of antimalarial sulphones and sulphonamides, is shown in Table 8.1.

The application of PCA to these data gave two principal components which explained 88 % and 8 % of the variance respectively. Relationships between variables may be seen by construction of a loadings plot, as shown in Figure 8.2; here the results of *Mycobacterium lufu* and *Escherichia coli* are seen to correspond to one another; the *Plasmodium berghei* data is quite separate.

Pictures such as this can be useful in predicting the likely specificity of compounds and can also give information (potentially) on their mechanism of action. When different test results are grouped very closely together it might be reasonable to suppose that inhibition of the same enzyme is involved, for example. A multivariate response set need not be restricted to data from the same type of biological test. Nendza and

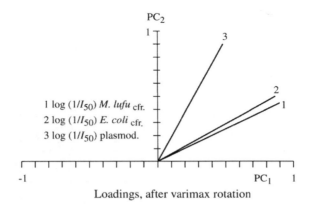

Figure 8.2 Loadings plot for three biological responses on the first two principal component axes (reproduced from ref. [1] with permission of Wiley-VCH).

Table 8.1 Observed biological activity of 2',4'-substituted 4-aminodiphenylsulfones determined in cell-free systems (columns 1–3, I_{50} [μmol/L]) and whole cell systems (columns 4–6, I_{25}, MIC [μmol/L]) of plasmodia and bacterial strains as indicated (reproduced from ref. [1] with permission of Wiley-VCH).

No.	Compound	P. berghei I_{50}	M. lufu I_{50}	E. coli I_{50}	E. coli I_{25}	M. lufu MIC	E. coli MIC
1	4'-NH$_2$(DDS)	12.41	1.20	34.36	7.857	0.17	16.00
2	4'-OCH$_3$	***	7.55	128.16	37.2777	30.38	109.00
3	4'-NO$_2$	***	31.03	212.74	39.546	10.78	5.60
4	4'-H	104.00	12.06	116.53	33.456	51.44	45.00
5	4'-OH	32.23	1.50	34.92	5.867	20.06	22.50
6	4'-Cl	***	12.99	***	***	59.76	45.00
7	4'-NHCOCH$_3$	33.26	7.01	75.65	21.406	10.33	45.00
8	4'-Br	***	9.69	***	***	***	33.75
9	4'-NHCH$_3$	26.70	1.29	46.21	10.740	1.90	90.00
10	4'-NHC$_2$H$_5$	***	2.75	41.71	16.140	0.90	64.00
11	4'-CH$_3$	48.00	8.08	89.84	47.221	16.17	***
12	4'-N(CH$_3$)$_2$	21.51	1.76	***	42.374	12.660	***
13	4'-COOCH$_3$	147.0	11.75	149.29	123.950	13.73	***
14	4'-COOH	***	3.60	74.24	867.200	***	***
15	4'-CONHNH$_2$	76.510	12.72	155.39	36.158	27.46	64.00

* The blank entries indicate a compound not tested or a missing parameter.

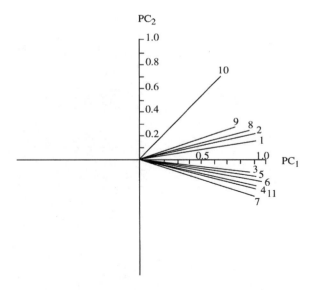

Figure 8.3 Loadings for eleven biological test systems on two principal components; the identity of the test systems is given in Table 8.3 (reproduced from ref. [2] with permission of Wiley-VCH).

Seydel [2] have reported results of the toxic effects of a set of phenols and aniline derivatives on three bacterial, four yeast, two algae, one protoplast, and one daphnia system (Table 8.2).

Loadings of these test systems on the first two principal components are shown in Table 8.3 and Figure 8.3.

The second principal component, which only explains 9 % of the variance in the set of 26 compounds, appears to be mostly made up from the inhibition of algae fluorescence data. All of the test systems have a positive loading with the first component but inspection of the loadings plot in Figure 8.3 shows that these variables fall into two groups according to the sign of their loadings with the second component. Scores may be calculated for each of the principal components for each of the compounds and these scores used as a composite measure of biological effect. The scores for PC_1 were found to correlate well with log k', a measure of hydrophobicity determined by HPLC, as shown in Figure 8.4. Thus, it may be expected that lipophilicity is correlated with the individual results of each test system.

Response data from both *in vitro* and *in vivo* test systems may be combined and analysed by PCA or FA. This is a particularly useful procedure since *in vitro* tests are often expected (or assumed!) to be good models for *in vivo* results. A straightforward test of the simple correlation

Table 8.2 Toxicity parameters of phenol and aniline derivatives (reproduced from ref. [2] with permission of Wiley-VCH).

	MIC $E.\ coli$	MIC M.169	I_{50} $E.\ coli$	I_{50} Saccharomyces cerevisiae	I_{50} Purin uptake	I_{50} ATPase	I_{50} DEF	I_{50} Rubisco	I_{50} Algae	I_{50} Fluor	ED_{100} Daphnia
1 Phenol	11	5.7	7.0	14	26	11	45	6.3	6.4	0.33	–
2 4–Cl–phenol	2.0	0.72	1.02	1.8	4.6	4.4	8.6	1.2	0.091	0.055	0.086
3 2,3–Cl$_2$–phenol	0.36	0.26	0.28	0.40	0.84	0.71	2.1	–	0.086	0.14	0.047
4 2,4–Cl$_2$–phenol	0.36	0.36	0.33	0.38	1.1	0.81	1.7	0.33	0.10	0.005	0.031
5 2,6–Cl$_2$–phenol	1.4	0.72	0.72	0.70	2.0	1.6	3.0	5.0	0.67	0.11	0.37
6 2,4,5–Cl$_3$–phenol	0.045	0.045	0.042	0.033	0.15	0.13	0.55	–	0.017	0.0038	0.013
7 2,4,6–Cl$_3$–phenol	0.36	0.13	0.19	0.060	0.15	0.44	0.52	0.93	0.011	0.36	0.056
8 2,3,4,5–Cl$_4$–phenol	0.016	0.011	0.045	0.010	0.014	0.023	0.12	–	–	–	–
9 Cl$_5$–phenol	0.13	0.0056	0.12	0.087	0.0035	0.014	0.11	0.009	0.0023	0.001	0.0075
10 2–Br–phenol	1.4	1.4	0.75	1.9	4.3	2.3	8.4	1.9	0.28	0.14	0.10
11 Aniline	23	11	26.2	23	72	50	105	36	0.67	0.41	4.1
12 2–Cl–aniline	5.7	4.1	2.2	5.4	13	10	16	4.3	0.45	0.42	0.28
13 3–Cl–aniline	4.1	2.0	2.7	3.6	13	8.6	16	1.6	0.25	0.18	0.39
14 4–Cl–aniline	2.9	0.51	3.0	2.8	17	11	18	2.2	0.017	0.009	0.78
15 2,4–Cl$_2$–aniline	2.0	0.72	0.38	1.1	2.3	1.6	3.0	2.6	0.11	0.031	0.062
16 2,6–Cl$_2$–aniline	5.7	8.2	0.62	1.2	3.4	1.7	4.0	21	0.48	0.63	0.12
17 2–Br–aniline	2.9	4.1	1.1	3.6	8.7	7.8	11	5.2	0.38	0.15	0.17
18 2,4–(NO$_2$)$_2$–aniline	1.0	0.51	0.41	1.5	0.11	0.80	–	0.02	0.0049	0.093	0.11
19 4–CH$_3$–aniline	8.2	4.1	12.0	11	23	22	50	31	0.13	0.079	4.7
20 2,4–(CH$_3$)$_2$–aniline	8.2	0.72	2.9	6.4	9.0	11	18	7.9	0.096	0.077	1.7

Table 8.3 Loadings of the biological test systems on the first two principal components (reproduced from ref. [2] with permission of Wiley-VCH).

	Test system	PC_1	PC_2
1	MIC *E. coli*	0.89	0.14
2	MIC M. 169	0.91	0.21
3	I_{50} *E. coli*	0.89	−0.13
4	I_{50} *Sacch. cerevisiae*	0.92	−0.24
5	I_{50} Purin	0.92	−0.16
6	I_{50} ATPase	0.95	−0.20
7	I_{50} DEF	0.90	−0.32
8	I_{50} Rubisco	0.88	0.24
9	I_{50} Algae	0.75	−0.27
10	I_{50} Fluorescence	0.65	0.70
11	I_{100} *Daphnia* (24 h)	0.89	−0.26

between any two experimental systems can of course be easily obtained but this may not reveal complex relationships existing in a response set. Figure 8.5 shows the results of a factor analysis of a combined set of *in vitro*, *in vivo*, and descriptor data, there being no reason why data sets should not be formed from a combination of dependent and independent data.

The factor plot shows that a calculated (Ke*-pred) and experimental (Ke*) measure of chemical reactivity fall close together, while in another part of factor space, experimental Ames test (STY) results (*in vitro*) and a predicted measure of mutagenicity (SA) are associated. Both of these

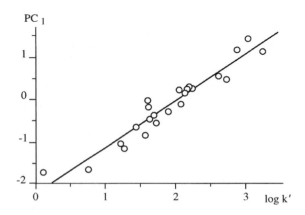

Figure 8.4 Plot of PC_1 scores versus log k' (reproduced from ref. [2] with permission of Wiley-VCH).

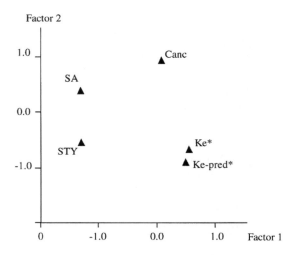

Figure 8.5 Loadings plot from a factor analysis of *in vivo* and *in vitro* tests, and measured and calculated physicochemical descriptors (reproduced from ref. [3] by permission of Oxford University Press).

associated sets of responses are separated from the *in vivo* measures of rat carcinogenicity (Canc), which they are expected to predict (at least to some extent). Another example of the use of factor analysis in the treatment of multiple response data involved the antitumour activity of platinum complexes against the tumour cell lines shown in Table 8.4 [4].

Three factors were extracted which explained 84 % of the variance of 52 complexes tested in these nine different cell lines. A plot of the

Table 8.4 Cell lines used in the testing of antitumour platinum complexes (reproduced from ref. [4] with permission of The Pharmaceutical Society of Japan).

Number on	Tumour cell lines	
Figure 8.6	Cell line	Origin
1	L1210	Mouse leukemia
2	P388	Mouse leukemia
3	LL	Mouse lung carcinoma
4	AH66	Rat hepatoma
5	AH66F	Rat hepatoma
6	HeLa S_3	Human cervical carcinoma
7	KB	Human nasopharyngeal carcinoma
8	HT-1197	Human bladder carcinoma
9	HT-1376	Human bladder carcinoma

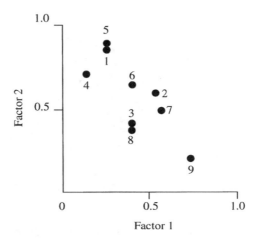

Figure 8.6 Loadings plot from a factor analysis of the activity of platinum complexes in a set of nine different tumour cell lines. Cell lines are identified by the numbers in Table 8.4 (reproduced from ref. [4] with permission of The Pharmaceutical Society of Japan).

rotated[1] factor loadings for the first two factors is shown in Figure 8.6 where it can be seen that the tests fall into four groups: AH66F, L1210, AH66; HeLa, P388, KB; HT-1197, LL; and HT-1376. Compounds that exhibit a given activity in one of these cell lines would be expected to show similar effects in another cell line from the same group, thus cutting down the need to test compounds in so many different cell lines.

The factor scores for the platinum complexes also present some interesting information. Figure 8.7 shows a plot of the factor 2 scores versus factor 1 scores where the points have been coded according to their activity against L1210 *in vivo*. Factor 2 appears to broadly classify the compounds in that high scores on factor 2 correspond to more active compounds; factor 1 does not appear to separate the complexes.

This plot indicates that the results obtained in the *in vitro* cell lines, as represented by the factor scores for the complexes, can be used as a predictive measure of *in vivo* activity. The factor scores can also be used as a simple single measure of 'anticarcinogenic' activity for use in other methods of analysis. The complexes were made up of carrier ligands and leaving groups and the activity contribution of each type of ligand or leaving group to the factor 2 scores was evaluated by the Free–Wilson method (Section 6.3.3). The results of this analysis are shown in

---

[1] Simplified by varimax rotation, see Section 4.2.

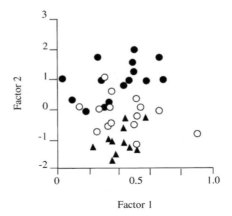

Figure 8.7   Scores plot of the first two factor axes for high activity (•), moderate activity (○), and low activity (▲) complexes against L1210 (reproduced from ref. [4] with permission of The Pharmaceutical Society of Japan).

Table 8.5 where the higher positive values indicate greater contribution to the factor scores, and hence *in vivo* activity.

Factor scores estimated by the Free–Wilson method are well correlated to the measured factor scores as shown in Figure 8.8. Factor scores or principal component scores can thus be used as new variables representing a set of multiple responses, just as they have been used as new variables representing a set of multiple descriptors. Section 8.5 describes

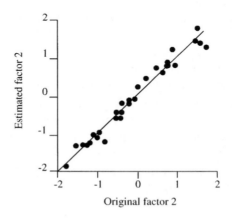

Figure 8.8   Plot of factor 2 scores estimated from a Free and Wilson analysis versus experimental factor 2 scores (reproduced from ref. [4] with permission of The Pharmaceutical Society of Japan).

Table 8.5 Contributions of carrier ligands and leaving groups to factor 2 scores (reproduced from ref. [4] with permission of The Pharmaceutical Society of Japan).

Carrier ligand	Contribution	Leaving group	Contribution
1-(Aminomethyl)cyclohexylamine	2.5694	Tetrachloro	0.0599
1-(Aminomethyl)cyclopentylamine	2.2497	Dichloro	0.0
1,1-Diethylethylenediamine	1.7729	Oxalato	−0.1479
1,2-Cyclohexanediamine	1.6738	Malonato	−0.2581
1,4-Butanediamine	0.7494	Sulfato	−0.2438
1,1-Dimethylethylenediamine	0.7370	2-Methylmalonato	−0.3393
3,4-Diaminotetrahydropyran	0.5084	Cyclobutane–1,1–dicarboxylato	−0.8787
Diamine	0.0	Dihydroxydichloro	−1.4111
N,N-dimethylethylenediamine	−0.08957		

the construction of regression-like models using multiple response and descriptor data.

8.3   CLUSTER ANALYSIS

We have already seen, in Section 5.4 on cluster analysis, the application of this method to a set of multiple response data (Figure 5.10). In this example the biological data consisted of 12 different *in vivo* assays in rats so the y (dependent) variable was a 40 (compounds) by 12 matrix. These compounds exert their pharmacological effects by binding to one or more of the neurotransmitter binding sites in the brain. Such binding may be characterized by *in vitro* binding experiments carried out on isolated tissues and a report byTesta *et al.* [5] lists data for the binding of 21 neuroleptic compounds to the following receptors:

α-noradrenergic, α1 and α2
β-noradrenergic, $\beta(1 + 2)$
dopaminergic, D1 and D2
serotoninergic, 5HT1 and 5HT2
muscarinic, $M(1 + 2)$
histaminic, H1
opioid
Ca^{2+} channel
serotonin uptake

A dendrogram showing the similarities between these compounds is given in Figure 8.9 where it can be seen that there are three main clusters. Cluster A, which is quite separate from the other clusters, contains the benzamide drugs. The compounds in cluster B, made up of two subgroups of different chemical classes, are characterized by higher affinity for D2 and 5HT2 receptors compared to D1, α1 and 5HT1 and have no activity at muscarinic receptors. Cluster C contains compounds with high α1 affinity, similar D1 and D2 affinity and a measurable affinity for muscarinic receptors (all other compounds have $-\log IC_{50} < 4$ for this receptor). Compounds shown in bold in Figure 8.9 were also present in the data set shown in Figure 5.10 and in some cases fall into similar clusters, e.g. fluphenazine and trifluperazine in cluster B1 are present in the same cluster in Figure 5.10. This shows the anticipated result that *in vitro* receptor binding data may be used to explain *in vivo* pharmacological results.

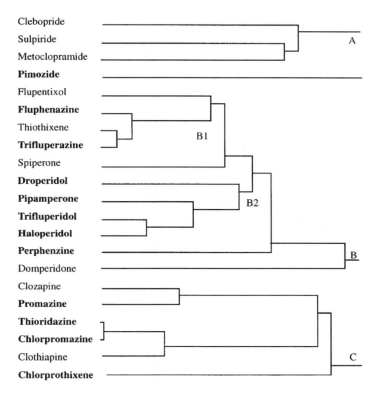

Figure 8.9 Dendrogram showing the associations between neuroleptic compounds binding to 14 receptors (reproduced from ref. [5] with kind permission of Springer Science + Business Media).

Cluster analysis can also be used to show relationships between variables, dependent or independent. Again, as was said in Section 5.4, a data set of n objects in a P-dimensional space can also be viewed as a set of P objects in an n-dimensional sample space. This is demonstrated in Figure 8.10 for the biological test results shown in Table 8.6 [6].

All of the test results are related to one another to some extent as shown by the correlation coefficients in the table and more graphically by the single large cluster in the dendrogram. This nicely illustrates the utility of a dendrogram since it is immediately obvious that $\log TD_{50}/ED_{50}$ is separated from the rest of the tests. This can be seen by inspection of the correlation matrix, for example, the bottom line ($\log TD_{50}/ED_{50}$) where the highest value is 0.494, but this is not as immediately apparent as the dendrogram. The dendrogram also clearly shows the unfortunate similarity between the dose required for therapeutic effect (ED_{50}) and the dose which shows toxicity (TD_{50}).

Table 8.6 Correlation matrix for biological test results for 41 nitro-9-aminoacridine derivatives (reproduced from ref. [6] with permission of Anticancer Drug Design).

	T1	T2	T3	T4	p LD_{50}	p ED_{50}	p TD_{50}	log LD_{50}/ED_{50}	log TD_{50}/ED_{50}
T1	1.000								
T2	0.781	1.000							
T3	0.724	0.684	1.000						
T4	0.800	0.846	0.686	1.000					
p LD_{50}	0.849	0.811	0.797	0.788	1.000				
p ED_{50}	0.847	0.754	0.721	0.896	0.877	1.000			
p TD_{50}	0.816	0.777	0.738	0.886	0.877	0.980	1.000		
log LD_{50}/ED_{50}	0.686	0.570	0.523	0.822	0.608	0.914	0.881	1.000	
log TD_{50}/ED_{50}	0.433	0.153	0.154	0.374	0.289	0.447	0.260	0.494	1.000

T1 - inhibition of *S. cerevisiae*
T2 - inhibition of germination of *L. sativum* sprouts
T3 - dehydrogenase activity inhibition in mouse tumour cells
T4 - inhibition of HeLa cell growth

p LD_{50} - measure of acute toxicity in mice
p ED_{50} - measure of antitumour activity in mice
p TD_{50} - measure of toxicity in tumour-bearing mice

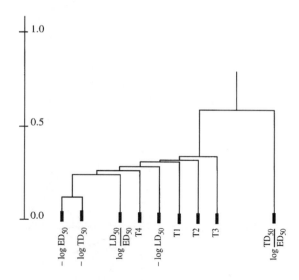

Figure 8.10 Dendrogram of the associations between the biological tests shown in the correlation matrix in Table 8.6 (reproduced from ref. [6] with permission of Anticancer Drug Design).

8.4 SPECTRAL MAP ANALYSIS

Principal component scores plots involve two principal components as the axes of the plot; similarly, principal component loadings plots make use of two principal components as the plot axes. Therefore it should be possible to plot both the scores and the loadings simultaneously on a plot, given the application of some suitable scaling. Such a plot is known as a biplot [7] and an example is shown in Figure 8.11 for the charge-transfer data set described in Section 7.3.1.

In the figure the diamonds represent the principal component scores for the individual rows (compounds) in the data set. These can obviously be labelled with their row numbers, coloured according to the dependent variable, and so on. The arrows represent the loadings of the named variables on the two principal component axes. The direction of the arrow from the origin indicates the sign of the loading and the length of the arrow is related to the magnitude of the loading. The loadings for the variables on these first two principal components are given in Table 7.7 and here it can be seen that μ_x and μ (shown as Mux and Mu on the figure) have positive loadings on PC2 whereas CMR, ClogP and E_{HOMO} have negative loadings. This is reflected in the directions of the corresponding arrows on the y-axis (PC2) in Figure 8.11. P3 and

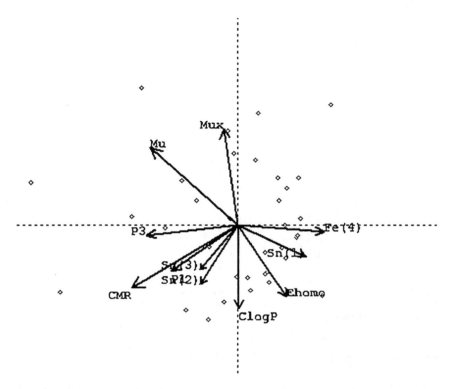

Figure 8.11 Biplot of the charge-transfer data produced using the program XLS-Biplot (http://gauss.upf.es/xls-biplot/).

CMR have positive loadings on PC1 in the table but in the figure are shown with negative loadings. This is just because of the direction that the program has arbitrarily chosen for this particular PC and in fact there is an option on the program to reverse the sign of either of the axes. Of course, as pointed out earlier, the 'direction' of a principal component has no particular significance, it is the relative signs and magnitude of the variable loadings on the PC's that are important. So, a biplot allows the analyst to simultaneously examine relationships between variables and samples in a single display and there are situations where this approach can be very useful.

A more spectacular example of biplots is spectral map analysis (SMA) which was briefly mentioned in Chapter 4 with an example of a spectral map shown in Figure 4.3. The reason for the development of this technique was the use of activity spectra to represent the activity of compounds in several different pharmacological (usually, but not necessarily) tests. An example of activity spectra for four α-agonists in six tests on rats is shown in Figure 8.12.

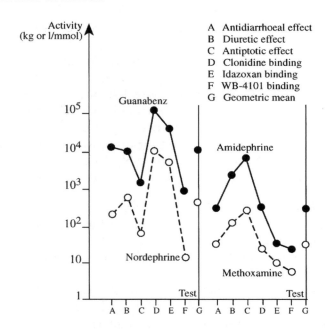

Figure 8.12 Activity spectra for four α-agonists (reproduced from ref. [8] with permission of Elsevier).

The three tests A, B, and C are *in vivo* antidiarrhoeal, diuretic, and antiptotic respectively; tests D, E, and F are *in vitro* clonidine binding [8]. The similarities between the two compounds in each pair can immediately be seen. The shape of their activity profiles, or spectra, demonstrate that guanabenz and nordephrine have similar selectivity for the six test results shown and simply differ in terms of their potency. Similarly, amidephrine and methoxamine have virtually identical spectra with amidephrine being the more active (the activity scale is log $1/C$ to cause a standard effect).

The distinction between activity and specificity has been likened by Lewi [8] to the difference between size and shape. As an example, he suggested the consideration of a table of measurements of widths and lengths of a collection of cats, tigers, and rhinoceroses. With respect to size, tigers compare well with rhinoceroses, but with respect to shape, tigers are classified with cats. In order to characterize shape it is necessary to compare width/length ratios with their geometric mean ratio. This is called a contrast and can be defined as the logarithm of an individual width/length ratio divided by the mean width/length ratio. From the table of animal measurements, a positive contrast would indicate a rhinoceros (width/length ratio greater than the mean width/length ratio), whereas a negative contrast could indicate a cat or a tiger. If the contrast is near zero, this may mean an overfed cat, a starving rhinoceros, or some error in the data.

For the compounds and tests represented by the activity spectra shown in Figure 8.12, a single major contrast allowed the correct classification to be made visually. However, when more than one important contrast is present in the data it becomes difficult, if not impossible, to identify these contrasts by simple inspection of activity spectra. This is where SMA comes in as it is a graphical method for displaying all of the contrasts between the various log ratios in a data table. The SMA process consists of the following steps.

- logarithmic transformation of the data;
- row centring of the data (subtraction of the mean activity of a compound in all tests);
- column centring of the data (subtraction of the mean activity of all compounds in one test);
- application of factor analysis to the doubly centred data;
- application of scaling[2] to allow both factor scores and loadings to be plotted on the same plot (called a biplot).

The end result of this procedure is the production of a biplot in which the similarities and differences between compounds, tests and compound/tests can be seen.

The activity spectra shown in Figure 8.12 come from a data set of 18 α-agonists tested in the six different tests. Application of SMA to this data set produced three factors which explained 48, 40, and 8 % respectively of the variance of contrasts. The factor scores and loadings were used to produce the biplot shown in Figure 8.13, the x and y axes representing the first and second factors, respectively.

At first sight it appears that this figure is horribly complicated but it does contain a great deal of information and the application of a few simple 'rules' does allow a relatively easy interpretation of the plot. The rules are as follows.

1. Circles represent compounds and squares represent tests. For example, oxymetazoline and clonidine binding in Figure 8.13.
2. Areas of circles and squares are proportional to the mean activity of the compounds and tests. For example, the mean activity of compounds in the antidiarrhoeal and diuretic tests are similar.

---

[2] Various scaling options can be applied to produce biplots, see ref. [8] for details.

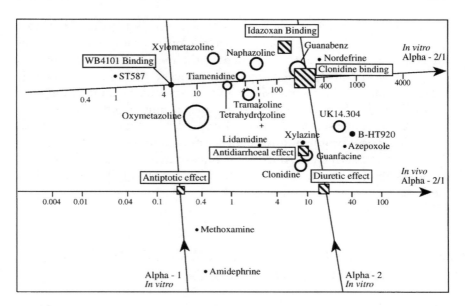

Figure 8.13 Spectral map of factor scores and loadings (reproduced from ref. [8] with permission of Elsevier).

3. The position of compounds and tests are defined by their scores and loadings on the first two factors. Where compounds are close together they have similar activity profiles; where tests are close together they give similar results for the same compounds; where compounds are close to tests they have high specificity for those tests. Compounds and tests that have little contrast lie close to the centre of the map. For example nordephrine and guanabenz are close together on Figure 8.13 and have similar activity profiles as shown in Figure 8.12. These compounds are close to the square symbol for clonidine binding (test D in Figure 8.12).

4. The third most significant factor is coded in the thickness of the contour around a symbol – a thick contour indicates that the symbol is above the plane of the plot, a thin contour that it lies below it.

5. Compounds and tests that are not represented in the space spanned by these three factors are represented by symbols with a broken line contour (none in this example).

6. An axis of contrast can be defined through any two squares representing tests.

There are many useful features of a data set that can be revealed in a spectral map such as this and we can see that it is consistent with the patterns shown in the activity spectra (Figure 8.12) since these two pairs of compounds are grouped together on the plot. The use of the thickness of symbol contours to denote the third factor is perhaps not very successful but modern computer graphics would allow easy display of such a plot in three dimensions. In this particular example, the third factor is probably not very important since it only describes 8 % of the variance of contrasts. SMA is clearly a useful method for the analysis of any chemical problem in which a set of compounds is subjected to a battery of tests which produce some quantitative measure of response. It offers the advantage of a simultaneous display of the relationships between both tests and compounds.

8.5 MODELS FOR MULTIVARIATE DEPENDENT AND INDEPENDENT DATA

Chapters 6 and 7 described the construction of regression models (MLR, PCR, PLS, and continuum regression) in which a single dependent variable was related to linear combinations of independent variables. Can these procedures be modified to include multiple dependent variables? One fairly obvious, perhaps trivial, way to take account of the information in two dependent variables is to use the difference between them or to take a ratio such as $\log TD_{50}/ED_{50}$ as seen in Section 8.3. Another way to take account of at least some of the information in a multivariate dependent set is to carry out PCA or FA on the data and use the resulting scores to construct regression models.

The biological activity data for the nitro-9-aminoacridines shown in Table 8.6 and Figure 8.10 were analysed by PCA to give two principal components explaining 81 % and 8 % of the variance in the set [6]. The loadings of the biological tests on these two principal components are shown in Table 8.7.

All of the tests appear to have a high positive loading on the first component and this PC was interpreted as being a measure of 'general biological activity'. The main variable loading onto PC_2 was the therapeutic index, although this loading is quite small (0.531). Thus the second PC might be interpreted as a measure of 'selectivity'. The compounds were described by log P and a number of topological descriptors, and regression equations were sought between the principal component scores and these physicochemical parameters. The first component

Table 8.7 Loadings of the two significant principal components after varimax rotation (reproduced from ref. [6] with permission of Anticancer Drug Design).

Biological tests[*]	Loadings for	
	PC_1	PC_2
T1	0.903	−0.124
T2	0.863	−0.252
T3	0.812	−0.386
T4	0.934	0.114
p LD_{50}	0.918	−0.252
p ED_{50}	0.971	0.195
p TD_{50}	0.967	0.151
log LD_{50}/ED_{50}	0.834	0.531

[*]The symbols of the biological tests have the same meaning as in Table 8.5.

was well described by a parabolic relationship with modified (see paper for details) log P values

$$PC1 = -0.25(\pm 0.03)\log P_*^2 - 0.80(\pm 0.08)\log P_* + 0.30(\pm 0.11) \quad (8.1)$$
$$n = 28 \qquad R = 0.97 \qquad s = 0.23 \qquad F = 201$$

The values in brackets are the standard errors of the regression coefficients and it should be noted that R is quoted not R^2 as is more usual for multiple regression equations. The second PC was less well described by a shape parameter based on molecular connectivity (^2K) and an indicator variable.

$$PC2 = 0.028(\pm 0.018)^2 K - 0.122(\pm 0.07) I_{N2N} - 0.124(\pm 0.101) \quad (8.2)$$
$$n = 28 \qquad R = 0.762 \qquad s = 0.07 \qquad F = 17.33$$

Equation (8.2) only describes 60 % of the variance in PC_2 and the high standard error for the shape descriptor term casts some doubt on the predictive ability of the equation. However, it is hoped that these two equations demonstrate the way in which regression models for multivariate dependent data can be generated by means of PCA.

Two alternative methods for the construction of regression type models for multivariate response sets are Partial Least Squares regression (PLS) and a technique known as *canonical correlation analysis* (CCA). PLS has already been described in Chapter 7 (7.3.2) as a regression method for constructing models between a single dependent variable and a set of descriptors but this technique can be more generally applied to a matrix (set) of dependents and a matrix of independent variables. A simple example may serve to illustrate this. Measurements of solubility (log$S$) and partition coefficient (log$P$) have been modelled for a set of 552 compounds using multiple linear regression applied to a set of 37 calculated molecular descriptors as the independent variables [9]. The following rather large regression equations were fitted for log$S$ with 28 and log$P$ with 26 significant descriptors:

$$\log S = \sum (a_i S_i) + 1.128 \tag{8.3}$$
$$n = 552, \quad R^2 = 0.783, \quad s = 0.753, \quad F = 68.18$$

$$\log P = \sum (a_i S_i) - 0.635 \tag{8.4}$$
$$n = 552, \quad R^2 = 0.870, \quad s = 0.645, \quad F = 134.8$$

Running the PLS regression routine [10] of the statistics package, R, on this data set gave rise to a model with 37 latent variables, each one containing contributions from all of the original variables. Such a high dimensional model is not necessary of course and the problem then is how to select the appropriate size for a useful model. Table 8.8 shows the percentage variance explained for the X set (descriptors) and the two responses with increasing numbers of latent variables in the model.

As can be seen from the table the variance explained steadily increases with increasing number of components, as expected, but there is no clear indication of where the model fitting should be stopped. As discussed earlier, this is one of the problems with fitting PLS models, the number of components to use, and there is no generally applicable test so the number of components to retain is a subjective judgement [10]. A plot of the Root Mean Squared Error of Prediction (RMSEP), however, is more informative. As can be seen from the plot (Figure 8.14) there is a distinct discontinuity at 4 components and thus it might be reasonable to stop the model at this point or shortly after.

The legend on the plot indicates that there should be two sets of values corresponding to CV and adjusted CV. This particular run was carried out with leave-one-out cross-validation which of course with a data set of this size has very little effect since only 1 point is omitted at a time from 552. The run was repeated with 10 randomly selected cross-validation segments, thus giving 10 % cross-validation results, and these are shown in the last 4 rows of Table 8.8. As can be seen from the table, the model gives an R^2 of 0.63 for log$S$ for 5 components and 0.69

Table 8.8 Percentage variance explained by number of components for the descriptors (X) and responses. R^2 and R^2_{CV} for the two responses.

	1 comps	2 comps	3 comps	4 comps	5 comps	6 comps	7 comps	8 comps
X	24.82	49.06	66.99	72.29	77.65	84.90	87.98	91.20
$\log S$	11.03	41.11	52.30	61.94	63.42	65.73	71.01	73.09
$\log P$	33.69	41.01	53.07	64.62	69.86	70.14	70.42	73.86
$\log S\ R^2$	0.11033	0.4111	0.5230	0.6194	0.6342	0.6573	0.710	0.731
R^2_{CV}	0.09094	0.3992	0.5107	0.5956	0.6164	0.6315	0.648	0.676
$\log P\ R^2$	0.3369	0.4101	0.5307	0.6462	0.6986	0.7014	0.7042	0.7386
R^2_{CV}	0.3180	0.3941	0.5130	0.6246	0.6805	0.6813	0.7086	0.7356

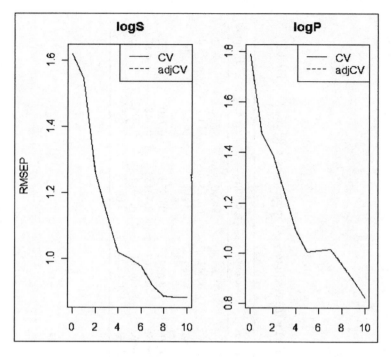

Figure 8.14 Plot of RMSEP *vs* the number of latent variables (a scree plot) for log$S$ and log$P$ in a combined PLS model.

for log$P$. Comparable results to the regression models can be obtained for larger numbers of components.

So, what about the other technique, canonical correlation analysis? CCA operates by the construction of a linear combination of q responses

$$W_1 = a_{1,1}Y_1 + a_{1,2}Y_2 + \ldots\ldots + a_{1,q}Y_q \qquad (8.5)$$

and a linear combination of p descriptors

$$Z_1 = b_{1,1}X_1 + b_{1,2}X_2 + \ldots\ldots + b_{1,p}X_p \qquad (8.6)$$

where the coefficients in Equations (8.5) and (8.6) are chosen so that the pairwise correlation between W_1 and Z_1 is as large as possible. W_1 and Z_1 are referred to as canonical variates and the correlation between them as the canonical correlation. Further pairs of canonical variates may be calculated until there are as many pairs of variates as the smaller of p and q. Important features of the canonical variates are listed below.

- They are generated in descending order of importance, i.e. canonical correlation coefficients decrease for successive pairs of variates.
- Successive pairs of variates are orthogonal to one another.

There should be a ring of familiarity about this description of canonical correlation when we compare it with PCA or FA and, indeed, CCA can be considered as a sort of joint PCA of two data matrices. The canonical variates are orthogonal to one another, as are principal components and factors, and they are generated in decreasing order of importance, although for CCA importance is judged by the correlation between canonical variates, not the amount of variance they explain which is the criterion used for principal components and factors. In this respect, CCA can also be seen to be akin to PLS since the latent variables in PLS are constructed so as to explain variance and maximize their covariance with a dependent variable. One other similarity which should be pointed out, which may not be immediately obvious, is the relationship between CCA and multiple linear regression (MLR). MLR normally involves a single response variable, Y and thus we can write

$$W_1 = Y \tag{8.7}$$

and this gives rise to one pair of canonical variates

$$Z_1 = b_{1,1} X_1 + b_{1,2} X_2 + \ldots\ldots + b_{1,p} X_p \tag{8.8}$$

where the coefficients are chosen to maximize the correlation between Z_1 and W_1 (Y). Equation (8.8) shows us that MLR can be viewed as a special case of canonical correlation analysis. Before moving on to an example of CCA, it is perhaps worth pointing out some obscuring jargon which is sometimes used to 'explain' the results of application of the technique. The canonical variates produced by linear combinations of the response set (W_1, W_2, and so on) are often called the nth canonical variate of the first set. Similarly, the linear combinations of the descriptor set (Z_1, Z_2, and so on) can be called the nth canonical variate of the second set. Thus, descriptions like 'the second canonical variate of the first set', and 'the first canonical variate of the second set' can confuse when what is meant is W_2 and Z_1 respectively.

Application of CCA to the log$S$ and log$P$ data set described above resulted in two pairs of canonical variates with canonical correlations of 0.92 and 0.75. These equate to correlation coefficients of 0.75 for log$S$ and 0.85 for log$P$ (see ref. [9] for details of how these correlation coefficients are extracted from the canonical correlations) thus CCA is giving similar results to the regression modelling and PLS, although with a much smaller number of latent variables than the PLS models.

Application of CCA to the neurotoxic effects of pyrethroid analogues has been reported by Livingstone and co-workers [11]. Factor analysis of a set of computed physicochemical properties for these compounds has been discussed in Section 5.3 of Chapter 5 (Table 5.9). Analysis of two *in vivo* responses, knockdown (KDA) and kill (KA), and eight *in vitro* responses by factor analysis led to the identification of three significant factors as shown in Table 8.9.

Factor 1 is associated with both killing and knockdown activity (factor loadings of -0.93 and -0.5 respectively) and several neurotoxicological responses, while factor 2 is mostly associated with knockdown. The third factor, although judged to be statistically significant, is almost entirely composed of the slope and intercept of the *in vitro* dose response curves and was judged not to have neurotoxicological significance. From this factor analysis of the combined *in vitro* and *in vivo* data, it was possible to identify three *in vitro* responses which had high association with the *in vivo* data, in itself a useful achievement since it is usually easier (and more accurate) to acquire *in vitro* data. The responses were the logarithm of the time to maximum frequency of action potentials (LTMF) which has a high loading on factor 1, the logarithm of the maximum burst frequency (LBF) which has a high loading on factor 2 and the logarithm of the minimum threshold concentration (MTC) which loads on to both factors. These three responses were used to select a subset of six physicochemical properties (see paper for details) which describe the pyrethroid analogues. Canonical correlation analysis of this set of three *in vitro* responses and six physicochemical descriptors gave rise to two pairs of significant canonical variates with correlations of 0.91 and 0.88 respectively. The first canonical variate had a high association with MTC and thus might be expected to be a good predictor for both knockdown and kill. The second canonical variate had a significant association with LBF and thus should model knockdown alone. An advantage of CCA is the simultaneous use of both response and descriptor data but this can also be a disadvantage. In order to make predictions for a new compound it is necessary to calculate values for its

Table 8.9 Rotated factor loadings for ten biological response variables (from ref. [11] with permission of Elsevier Science).

	Killing activity (KA)	Time onset	Time to maximum frequency (LTMF)	Minimum threshold conc. (MTC)	Max. freq. of AP	Time to block	Slope	Intercept	Max. burst freq. (LBF)	Knock-down (KDA)
F1	−0.93	0.84	0.88	0.71	0.74	0.79	−0.16	0.21	0.06	−0.50
F2	0.28	−0.10	−0.29	−0.57	0.39	0.04	0.23	−0.18	0.90	0.72
F3	0.01	0.15	−0.10	0.00	−0.31	0.10	0.93	0.94	−0.04	0.16

* *In vivo* data (KA and KDA) measured using *Musca domestica l.*, *in vitro* data measured using isolated haltere nerves.

physicochemical properties and then to combine these using the coefficients from the canonical correlation to produce descriptor scores (scores for the second set). These scores are equated by the canonical correlation to response scores (scores for the first set) but of course the scores are linear combinations of the individual responses. In order to predict an individual response for a new compound, it is necessary to obtain measured values for the other responses. This may be advantageous if one of the response scores is difficult or expensive to measure, for example, an *in vivo* response. The ability to predict *in vivo* responses from *in vitro* data may be a significant advantage in compound design. It may also be possible to make an estimate for an individual response by making assumptions about the values of the remaining response variables. The extra complexity of an approach such as CCA may be a disadvantage but it also offers a number of advantages.

8.6   SUMMARY

This chapter has shown how multivariate dependent data, from multiple experiments or multiple results from one experiment, may be analysed by a variety of methods. The output from these analyses should be consistent with the results of the analysis of individual variables and in some circumstances may provide information that is not available from consideration of individual results. In this respect the multivariate treatment of dependent data offers the same advantages as the multivariate treatment of independent data. The simultaneous multivariate analysis of response and descriptor data may also be advantageous but does suffer from complexity in prediction.

In this chapter the following points were covered:

1. how PCA and factor analysis can be used to examine the relationships between different dependent (test) data;
2. modelling combined responses using PCA or FA to produce summary dependent variables;
3. simultaneous examination of dependent and independent data using PCA, FA and biplots/spectral maps;
4. the use of cluster analysis with multiple dependent variables;
5. modelling multiple dependent data with canonical correlation analysis and PLS.

REFERENCES

[1] Wiese, M., Seydel, J.K., Pieper, H., Kruger, G., Noll, K.R., and Keck, J. (1987). *Quantitative Structure–Activity Relationships*, **6**, 164–72.

[2] Nendza, M. and Seydel, J.K. (1988). *Quantitative Structure–Activity Relationships*, **7**, 165–74.

[3] Benigni, R., Cotta-Ramusino, M., Andreoli, C., and Giuliani, A. (1993). *Carcinogenesis*, **13**, 547–53.

[4] Kuramochi, H., Motegi, A., Maruyama, S., *et al.* (1990). *Chemical and Pharmaceutical Bulletin*, **38**, 123–7.

[5] Testa, R., Abbiati, G., Ceserani, R., *et al.* (1989). *Pharmaceutical Research*, **6**, 571–7.

[6] Mazerska, Z., Mazerska, J., and Ledochowski, A. (1990). *Anti-cancer Drug Design*, **5**, 169–87.

[7] Gabriel, K.R. (1971). *Biometrika*, **58**, 453–67.

[8] Lewi, P.J. (1989). *Chemometrics and Intelligent Laboratory Systems*, **5**, 105–16.

[9] Livingstone, D.J., Ford, M.G., Huuskonen, J.J. and Salt, D.W. (2001). *Journal of Computer-Aided Molecular Design*, **15**, 741–52.

[10] Mevik, B.-H. and Wehrens, R. (2007). *Journal of Statistical Software*, **18**, 1–24.

[11] Livingstone, D.J., Ford, M.G., and Buckley, D.S. (1988). In *Neurotox 88: Molecular Basis of Drug and Pesticide Action*, G.G. Lunt (Ed.), pp. 483–95. Elsevier, Amsterdam.

9

Artificial Intelligence and Friends

Points covered in this chapter

- Expert systems
- Log P and toxicity prediction
- Reaction routes and chemical structure prediction
- Artificial neural networks
- Network interrogation
- Rule induction
- Genetic algorithms
- Consensus models

PREAMBLE

This chapter is mostly concerned with techniques which can be broadly classified as artificial intelligence. A couple of other topics which are important in data analysis are also discussed here since they don't logically or easily fit elsewhere in this book. If artificial intelligence can be thought of as mimicking biological systems then so can genetic methods and hence the chapter title of artificial intelligence and friends. Consensus modelling doesn't fit into this analogy so easily!

A Practical Guide to Scientific Data Analysis   David Livingstone
© 2009 John Wiley & Sons, Ltd

9.1 INTRODUCTION

The workings of the human mind have long held a fascination for man, and we have constantly sought to explain how processes such as memory and reasoning operate, how the senses are connected to the brain, what different physical parts of the brain do, and so on. Almost inevitable consequences of this interest in our own minds are attempts to construct devices which will imitate some or all of the functions of the brain, if not artificial 'life' then at least artificial intelligence. It might be thought that we have already achieved this goal when some of the awesome computing tasks, such as weather forecasting, that are now carried out quite routinely (and surprisingly accurately) are considered. Nowadays, for example, even a simple electrical appliance like the humble toaster is likely to contain a microchip 'brain'. Computers, of course, have revolutionized artificial intelligence (AI) research, so much so that devices are now being built which are models, albeit limited, of the physical organization and 'wiring' of the brain. These systems are known as artificial neural networks (ANN) and they have proved to be so remarkably successful that they have found application in a very diverse set of fields as shown in Table 9.1. The use of ANN in the analysis of chemical data is discussed in Section 9.3.

AI research has already provided the concepts of supervised and unsupervised learning to data analysis, and these have proved useful in the classification of analytical methods and to alert us to the potential danger of chance effects. But what of the application of AI techniques themselves to the analysis of chemical data? The linear learning machine, or discriminant analysis (see Section 7.2.1) is an AI method that is used in data analysis, but perhaps the most widely used AI technique is a method called expert systems. There are various flavours of expert systems and

Table 9.1 Applications of artificial neural networks.

'Reasoning'*	Process control
Verification of handwriting on cheques	Traffic control on underground stations
Identification of faces for a security system	Control of a magnetic torus for nuclear fusion
Credit ratings	Control of a chemical plant
Stock market forecasting	Control of a nuclear reactor
Drug detection at airports	
Grading pork for fat content	

*Including pattern recognition.

some authors apply the term to models, e.g. regression models, which have been derived from a particular set of data. In this chapter, the term expert system is used only to mean a procedure which has been created by some human expert or panel of experts. The expert systems described here do not necessarily have a role in data analysis but they do have a role in chemistry, for example, to calculate octanol/water partition coefficients and predict toxicity. The next section of this chapter deals with expert systems and the following section contains miscellaneous examples of the use of AI methods in chemistry.

9.2 EXPERT SYSTEMS

The heart of any expert system consists of a set of rules, sometimes referred to as a rule base or knowledge base, which has been put together by 'experts'. The question of course arises of how to define (or find) the experts, but for the purposes of this discussion an expert is any human who has an opinion on the particular problem to be solved! Expert systems are usually, but not necessarily, implemented on a computer (see, for example, the structural alert system later in this section).

A simple example from the FOSSIL (Frame Orientated System for Spectroscopic Inductive Learning) system, which aims to identify chemicals from spectral data, may illustrate the expert system approach [1]. This system contains, in its knowledge base, information about NMR spectra, infrared spectra, mass spectra, and ultraviolet spectra along with spectral heuristics (most suitable technique for assignment of particular structural properties, etc.) and molecular structures, functional groups, etc. One of the rules in the infrared section concerns the assignment of a methyl stretch – are there peaks in the IR spectrum in the region 2950–2975cm^{-1} and 2860–2885cm^{-1}? This information may be obtained from the user of the system by a prompt (the program asks a question) or might come from the automatic interpretation of a spectrum. A positive answer to this question would indicate the presence of a methyl group, and it is easy to see how similar rules in this and other parts of the knowledge base would allow assignment of various structural features. Some spectral information, of course, may indicate the presence of several alternative molecular features and here it is necessary to construct logical queries using operators such as IF, AND, NOT, THEN, ELSE, etc. The information required to satisfy such a query may be contained entirely in one spectrum or may need to come from other sources such as NMR and UV. Once the molecular features have been recognized, the construction of a molecule from them is not a trivial task since there will be many ways in which such fragments can be put together. A commonly adopted approach to this type of problem in expert systems is the construction of a decision tree which contains a number of connected nodes. Each node represents a logical question which has two or more answers and which dictates how the expert system proceeds with a query, in this example, the construction of a molecule from the identified features. Further discussion of expert systems can be found in the book by Cartwright [2] and the review by Jakus [3].

9.2.1 Log P Calculation

One of the earliest applications of expert systems in the field of QSAR was the development of calculation schemes for octanol/water partition coefficients. Although the early work with π constants had shown that they were more or less additive (see Chapter 10), a number of anomalies had been identified. In addition, in order to calculate log P values from π constants it is necessary to have a measured log P for the parent and this, of course, is often unavailable. One approach to the question of how to calculate log P from chemical structure is to analyse a large number of measured log P values so as to determine the average contribution of particular chemical fragments [4]. The fragment contributions constitute the rules of the expert system, extra rules being supplied in the form of correction factors. Operation of this expert system consists of the following few simple steps.

- Break down the chemical structure into fragments that are present in the fragment table.
- Identify any correction factors that are needed.
- Add together the fragment values and apply the necessary correction factors to obtain a calculated log P.

Table 9.2 gives an example of some of the fragment contributions for this method; interestingly the correction factors always appeared to adopt the same value (0.28) or multiples of it. This was originally given the perhaps unfortunate name 'the magic constant'.

The Rekker system of log P calculation was based on a statistical analysis of a large number of measured partition coefficients and can thus be called a reductionist approach. An alternative procedure was proposed

Table 9.2 Some fragment constants and correction factors for the Rekker log P prediction method[a] (reproduced from ref. [5] copyright (1979) Elsevier).

Fragment[b]	Value	Fragment	Value
Br (Al)	0.249	COOH (Al)	−0.938
Br (Ar)	1.116	COOH (Ar)	−0.071
Cl (Al)	0.057	$CONH_2$ (Al)	−1.975
Cl (Ar)	0.924	$CONH_2$ (Ar)	−1.108
NO_2 (Al)	−0.920	CH_3	0.701
NO_2 (Ar)	−0.053	C_6H_5	1.840
OH (Al)	−1.470	pyridinyl	0.520
OH (Ar)	−0.314	indolyl	1.884

Correction factors	
Type	**Value**
Proximity effect – 2C separation	2×0.28
Proximity effect – 1C separation	3×0.28
H attached to a negative group	0.27
Ar–Ar conjugation	0.31
Proximity effects	
(experimental – summation of component fragments)	
–COOH	8×0.28
$–CONH_2$	4×0.28
$–NHCONH_2$	11×0.28

[a]This is only a small part of the scheme reported by Rekker and de Kort [5].
[b]The symbol in brackets denotes aliphatic (Al) or aromatic (Ar).

by Hansch and Leo [6] which involved a small number of 'fundamental' fragment values derived from very accurate partition coefficient measurements of a relatively small number of compounds. This technique, which can be viewed as a constructionist approach, requires a larger number of correction factors as shown in Table 9.3.

In both cases the procedure for the calculation of a log P value is the same; a compound is broken down into the appropriate fragments, correction factors are identified, and the fragments and correction factor values are summed up. Figure 9.1 illustrates the process for three different compounds where it can be seen that both methods can give quite comparable results which are in good agreement with the experimental values. This, of course, is not always the case; for some types of compounds, as shown in the figure, the Rekker method may give better estimates than the Hansch and Leo approach, and vice versa for other sets of compounds. A comparison of the approaches concludes, perhaps unsurprisingly, that neither can be said to be 'best' [7] and it is

Table 9.3 Some fragment values and correction factors for the Hansch and Leo log P prediction system[a] (reproduced from ref. [6] with permission of John Wiley & Sons, Inc.).

Fragment	Value[b]		
	f	f^{ϕ}	$f^{\phi\phi}$
Br	0.20	1.09	
Cl	0.06	0.94	
NO$_2$	−1.16	−0.03	
OH	−1.64	−0.44	
COOH	−1.11	−0.03	
CONH$_2$	−2.18	−1.26	
–O–	−1.82	−0.61	0.53
–NH–	−2.15	−1.03	−0.09
–CONH–	−2.71	−1.81	−1.06
–CO$_2$–	−1.49	−0.56	−0.09

Correction factors	
Type	Value
Normal double bond	−0.55
Conjugate to ϕ	−0.42
Chain single bond	−0.12 (proportional to length)
Intramolecular H–bond	0.60 (for nitrogen)
Intramolecular H–bond	1.0 (for oxygen)

[a]From the scheme (Tables IV-1a and IV-1b reported by Hansch and Leo [6]).
[b]Value given for aliphatic (f), aromatic (f^{ϕ}), and for the fragment between two aromatic system ($f^{\phi\phi}$).

prudent to always compare predictions with measured values whenever possible.

In these two examples, the knowledge base or rule base of the expert system consists of the fragment values and correction factors, along with any associated rules for breaking down a compound into appropriate fragments. Although the numerical values for the fragments and factors have been derived from experimental data (log P measurements) it has required a human expert to create the overall calculation scheme. This can be seen particularly clearly for the Hansch and Leo system which, because of the small number of fragments in the scheme, requires a variety of different types of correction factors to account for the way that different fragments influence one another. Devising the correction factors and the rules for their application has required the greatest contribution of human expertise in these two systems. One major problem in the application of any expert system to chemical structures is the question of how to break down compounds into fragments that will be recognized

$$HO - CH_2 - COOH \qquad\qquad \text{Measured log } P = -1.11$$

Rekker:

$$\text{Log } P = f_{OH,al} + f_{CH_2} + f_{COOH, al} + 3CM$$
$$= -1.022 \qquad\qquad \Delta = 0.09$$

Hansch and Leo:

$$\text{Log } P = f_{OH} + f_{CH_2} + f_{COOH} + (2\text{-}1)\, F_b + F_{P1}$$
$$= -1.06 \qquad\qquad \Delta = 0.05$$

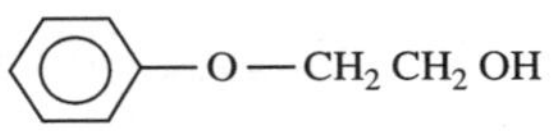

$$\text{Measured log } P = 1.16$$

where the structure is a phenyl ring bonded to $-O - CH_2\, CH_2\, OH$

Rekker:

$$\text{Log } P = f_{C_6H_5} + f_{O,or} + f_{CH_2} + f_{OH, al} + 2CM$$
$$= -1.55 \qquad\qquad \Delta = 0.39$$

Hansch and Leo:

$$\text{Log } P = f_{C_6H_5} + f_{O}{}^{\phi} + 2f_{CH_2} + f_{OH} + (4\text{-}1)\, F_b + F_{P2}$$
$$= -1.20 \qquad\qquad \Delta = 0.04$$

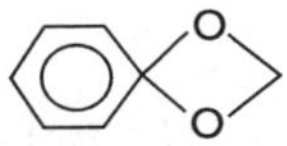

$$\text{Measured log } P = 2.08$$

Rekker:

$$\text{Log } P = f_{C_6H_4} + 2f_{O,or} + f_{CH_2} + 3CM$$
$$= 2.17 \qquad\qquad \Delta = 0.09$$

Hansch and Leo:

$$\text{Log } P = 4f_{CH}{}^{\phi} + 2f_{C}{}^{\phi} + 2f_{O}{}^{\phi} + f_{CH_2} + 3\, F_b + F_{P1} + F_{P2}{}^{\phi}$$
$$= -1.34 \qquad\qquad \Delta = -0.74$$

Figure 9.1 A comparison of the Rekker and the Hansch and Leo log P calculation for three molecules (reproduced from ref. [7] copyright (1982) Elsevier).

by the system. The problem lies not so much in the process of creating fragments but in deciding which are the 'correct' fragments. Although the rules for creating fragments are a necessary part of such expert systems, it is possible, particularly for large molecules, to create different sets of fragments which still conform to the rules. If the expert system has been carefully created, these different sets of fragments (and correction factors) may yield the same answer, but it is disconcerting to find two (or more) ways to do the same job. This was a particular problem when the schemes were first created and all calculations were carried out manually, but more recently they have been implemented in computer

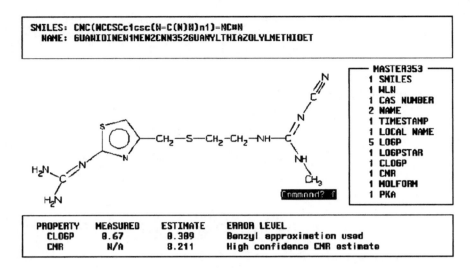

```
SMILES: CNC(NCCSCc1csc(N=C(N)N)n1)=NC#N
NAME:   GUANIDINEN1MEN2CNN3S2GUANYLTHIAZOLYLMETHIOET
```

	MASTER353
	1 SMILES
	1 WLN
	1 CAS NUMBER
	2 NAME
	1 TIMESTAMP
	1 LOCAL NAME
	5 LOGP
	1 LOGPSTAR
	1 CLOGP
	1 CMR
	1 MOLFORM
	1 PKA

Command? f

PROPERTY	MEASURED	ESTIMATE	ERROR LEVEL
CLOGP	0.67	0.389	Benzyl approximation used
CMR	N/A	8.211	High confidence CMR estimate

Figure 9.2 Computer screen for input of the compound tiotidine to the program UDRIVE (reproduced with permission of Daylight Chemical Information Systems Inc.).

systems. The Hansch and Leo system is probably the most widely used, and is available commercially from Daylight Chemical Information Systems (www.daylight.com). Of course, a computer implementation of a chemical expert system requires some means by which chemical structure information can be passed to the computer program. Unless the particular expert system simply requires a molecular formula (or some other atom count), this means that it is necessary to provide two- or three-dimensional information to the program. There are a variety of ways in which two- and three-dimensional chemical information can be stored and processed by computers; the input system used by the Hansch and Leo expert system of log P calculation (CLOGP) is known as SMILES (Simplified Molecular Input Line Entry System). The SMILES coding scheme is so elegantly simple and easy to learn (see Box 9.1) that it has become used as an input system for several other chemical calculation programs. Figure 9.2 shows the input screen for one means of access to the CLOGP program[1] for the compound tiotidine, an H_2 receptor antagonist used for the control of gastric acid secretion.

---

[1] This is the program UDRIVE which provides access to CLOGP and CMR calculation algorithms, as well as the database routine THOR running on DEC VAX machines (once a very popular scientific computer, defunct since 2005).

Box 9.1   SMILES - line entry for chemical structure

The SMILES (Simplified Molecular Input Line Entry System) system requires only four basic rules to encode almost all organic structures in their normal valence states. These rules are as follows.

1. Atoms in the 'organic' subset (B, C, N, O, P, S, F, Cl, Br, and I) are represented by their atomic symbols with hydrogens (to fill normal valency) implied. Thus,

   C - methane (CH_4)
   N - ammonia (NH_3)
   O - water (H_2O)
   Cl - hydrogen chloride (HCl)

   Atoms in aromatic rings are specified by lower case letters, e.g. normal carbon C, aromatic carbon c.

2. Bonds are represented by −, =, and # for single, double, and triple bonds respectively. Single bonds are implied and thus the − symbol is usually omitted, but the double and triple bonds must be specified. Thus,

CC	ethane (CH_3CH_3)
CCO	ethanol (CH_3CH_2OH)
C=C	ethylene ($CH_2{=}CH_2$)
O=C=O	carbon dioxide (CO_2)
C=O	formaldehyde (CH_2O)
C#N	hydrogen cyanide (HCN)

3. Branches are specified by enclosure of the branch within brackets, and these brackets can be nested or stacked to indicate further branching.

2-propylamine	$CH_3CH(NH_2)CH_3$	CC(N)C
isobutyric acid	$CH_3CH(CH_3)C(=O)OH$	CC(C)C(=O)O
3-isopropyl-1-hexene	$CH_2CHCH(CH(CH_3)_2)CH_2CH_2CH_3$	C=CC(C(C)C)CCC

4. Cyclic structures are represented by breaking one single or aromatic bond in each ring and numbering the atoms on either side of the bond to indicate it. This is shown for several different rings in the figure. A single atom may have more than one ring closure; different ring closures are indicated by different numbers (the digits 1–9 are allowed and can be reused after closure of that ring bond).

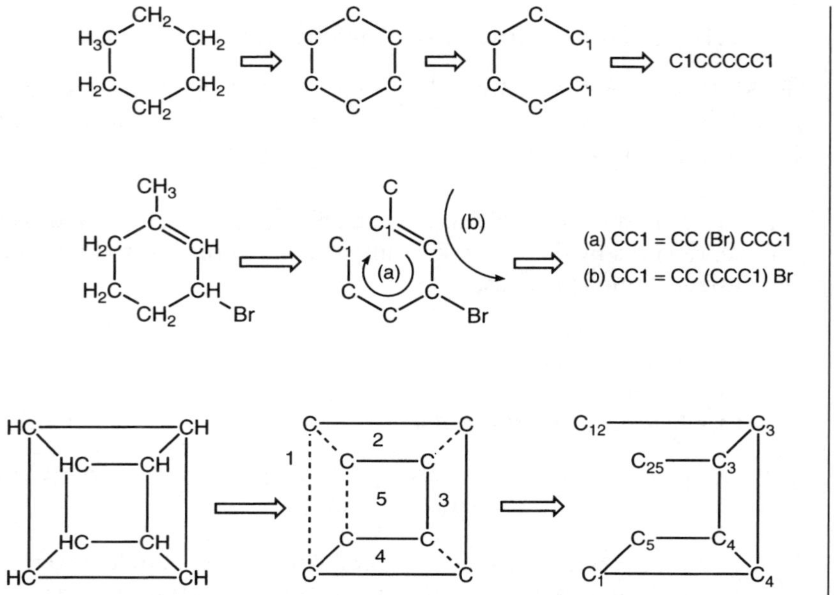

(a) CC1 = CC (Br) CCC1

(b) CC1 = CC (CCC1) Br

Generation of SMILES for cubane: C12C3C4C1C5C4C3C25

These simple rules allow very rapid encoding of most chemical structures and need only a few simple additions to cope with other atoms, charges, isomers, etc. Specification of atoms not in the 'organic' subset, for example, is coded by use of an atomic symbol within square brackets. An extensive description of the SMILES system is given by Weininger and Weininger [8].

One of the most attractive features of the SMILES structure generation algorithms is that is does not matter where a SMILES string begins, if the coding is correct the corresponding structure will be produced. This is in marked contrast to other linear chemical structure coding schemes where order is important and there are complex rules to decide where to start. The following are all valid SMILES for 6-hydroxy-1, 4-hexadiene.

$CH_2{=}CH{-}CH_2{-}CH{=}CH{-}CH_2OH$ C=CCC=CCO
                                                  C(C=C)C=CCO
                                                  OCC=CCC=C

Although the ordering of coding of a SMILES string does not matter for the input of structures, it does have an effect on the efficiency of storing and subsequent searching of a collection of compounds. The order of coding of structures also has implications for the generation of rules for chemical expert systems which, for example, predict three-dimensional chemical structure. By application of a set of ordering rules, it is possible to produce a unique SMILES string for any given structure. This is achieved in the Daylight software by the use of two separate algorithms.

The SMILES string CNC(NCCSCc1csc(N=C(N)N)n1)=NC#N is shown at the top of the screen with a two-dimensional representation of the compound in the box. A calculated value for log P (0.389) and molar refractivity (8.211) appear at the bottom of the screen and, in this case, a measured log P value (0.67) which has been retrieved from the Pomona College Medicinal Chemistry Database. The Daylight software provides access to a variety of chemical databases in addition to the CLOGP and CMR calculation routines (see Daylight website). The box on the right-hand side of Figure 9.2 reports a summary of the types of data which are held in the database which the program is currently connected to, in this case the Pomona College Master 353 data collection. Figure 9.3 shows

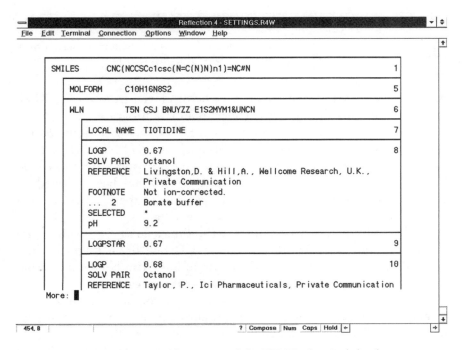

Figure 9.3 Computer screen for access of the THOR chemical database program from UDRIVE (reproduced with permission of Daylight Chemical Information Systems Inc.).

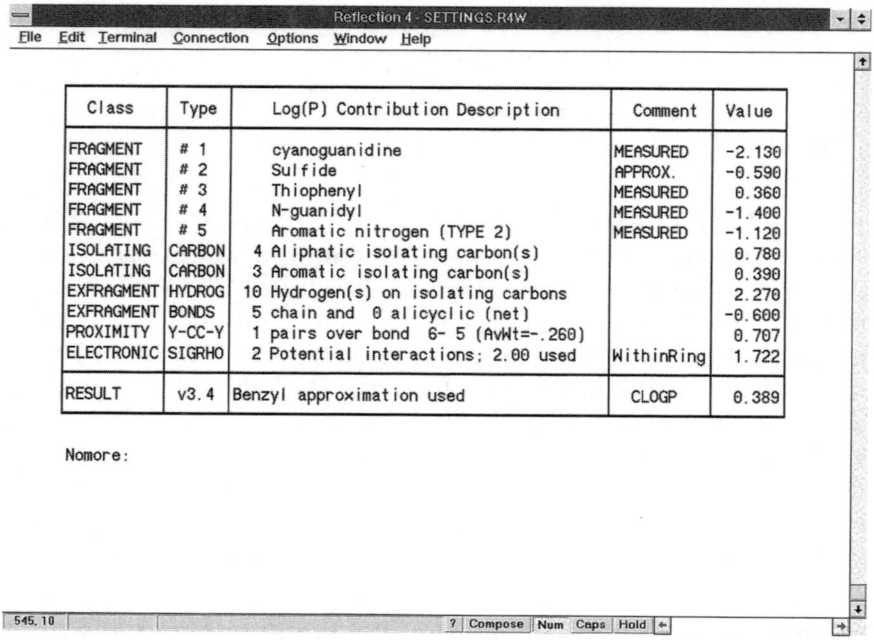

```
══                          Reflection 4 - SETTINGS.R4W                    ▼ ♦
File  Edit  Terminal  Connection  Options  Window  Help
                                                                            ▲
```

Class	Type	Log(P) Contribution Description	Comment	Value
FRAGMENT	# 1	cyanoguanidine	MEASURED	-2.130
FRAGMENT	# 2	Sulfide	APPROX.	-0.590
FRAGMENT	# 3	Thiophenyl	MEASURED	0.360
FRAGMENT	# 4	N-guanidyl	MEASURED	-1.400
FRAGMENT	# 5	Aromatic nitrogen (TYPE 2)	MEASURED	-1.120
ISOLATING	CARBON	4 Aliphatic isolating carbon(s)		0.780
ISOLATING	CARBON	3 Aromatic isolating carbon(s)		0.390
EXFRAGMENT	HYDROG	10 Hydrogen(s) on isolating carbons		2.270
EXFRAGMENT	BONDS	5 chain and 0 alicyclic (net)		-0.600
PROXIMITY	Y-CC-Y	1 pairs over bond 6- 5 (AvWt=-.260)		0.707
ELECTRONIC	SIGRHO	2 Potential interactions; 2.00 used	WithinRing	1.722
RESULT	v3.4	Benzyl approximation used	CLOGP	0.389

```
Nomore:



545.10                                       ?  Compose  Num  Caps  Hold  ←        ➡
```

Figure 9.4   Details of the calculation of log P for tiotidine (computer screen from UDRIVE) (reproduced with permission of Daylight Chemical Information Systems Inc.).

how the data for this compound may be displayed from the UDRIVE menu by accessing the THOR chemical database program.

The database page contains the SMILES string for tiotidine as the root of the data tree, the molecular formula and a WLN string,[2] a local name for the compound, and then experimental values for log P and pK_a in some cases. The measured value of 0.67 has been selected by the database constructors as a 'best' value, called a log P^*. The Pomona College database stores partition coefficient values for octanol/water and other solvent systems (in 2009, 61 000 measured log P values and 13 900 pK_a values for 55 000 compounds). Details of the fragments and correction factors used in the calculation of log P can be obtained from the program as shown in Figure 9.4. In this case the calculation made use of one approximated fragment as shown by the comments alongside the fragments and factors; the degree of certainty can be seen by comparison of the predicted with the measured value ($\Delta \log P = -0.28$).

---

[2] WLN (Wiswesser line notation) is another line notation system for chemical structures which has seen widespread use.

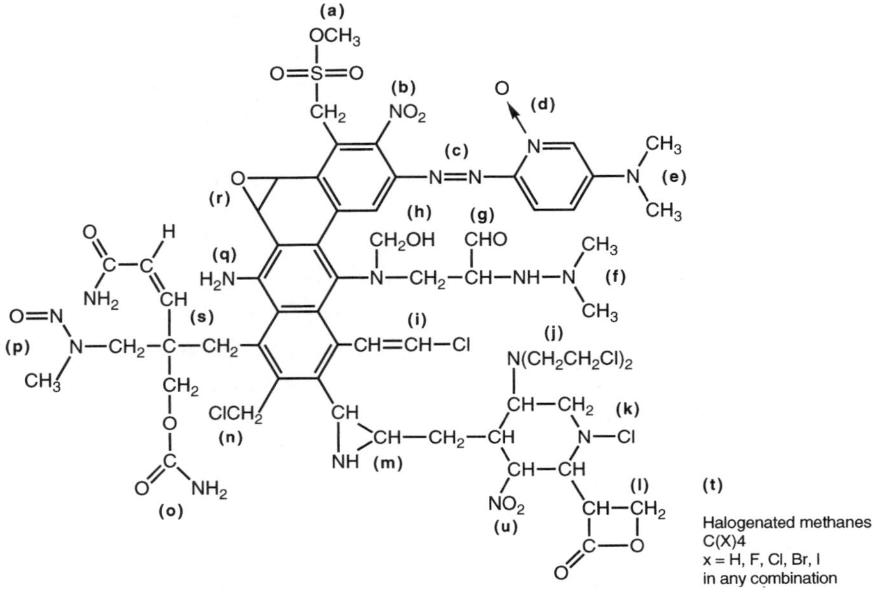

Figure 9.5 A supermolecule for the assessment of structural alerts for toxicity (reproduced from ref. [10] with permission of Elsevier).

9.2.2 Toxicity Prediction

The structural alert model for carcinogenicity first proposed by Ashby [9] and later modified by Ashby and Tennant [10] is a good example of a manual expert system in chemistry. This scheme was created by the recognition of common substructures which occur in compounds which have shown positive in an *in vitro* test for mutagenicity, the well-known Ames salmonella test. Putting together these substructures has allowed the creation of a 'supermolecule' as shown in Figure 9.5.

Prediction of the likelihood of mutagenicity for any new compound is achieved by simple comparison of the new structure with the alerts present in the supermolecule; common sense suggests that the greater the number of alerts, then the higher the likelihood of mutagenicity. The structural alert system has been shown to be successful for the prediction of mutagenicity; for example, in a test of 301 chemicals [11] almost 80 % of alerting compounds were mutagenic compared with 30 % of non-alerting compounds. Unfortunately, compounds may be carcinogenic due to mechanisms other than mutagenicity (thought to be caused by reaction with a nucleophilic site in DNA). The correlation between mutagenicity and carcinogenicity (measured in two rodent species) was low, suggesting that structural alerts are useful but nondefinitive indicators of potential carcinogenic activity [11]. This is a nice example of the strengths and limitations of expert systems; if the experts' knowledge is well coded (in the rules) and used correctly, the system will make good predictions. The expert system, however, can only predict what it really knows about; if a compound is carcinogenic because it is mutagenic then all is well.

The DEREK system (Deductive Estimation of Risk from Existing Knowledge) is a computer-based expert system for the prediction of toxicity [12]. This program uses the LHASA synthesis planning program (see Section 9.2.3) as its foundation for the input of chemical structures and the processing of chemical substructures. DEREK makes predictions of toxicity by the recognition of toxic fragments, toxicophores, defined by the rules present in a rule base created by human experts. At present, the rule base is being expanded by a collaborative effort involving pharmaceutical, agrochemical, and other chemical companies, as well as government organizations. The collaborative exercise involves a committee which considers any new rules that are presented for inclusion in the rule base. Rules are written in the PATRAN language of LHASA, and when a rule is activated, due to the presence of a toxic fragment, a different language, CHMTRN, is used to consider the rest of the structure and the environment of the toxicophore. Toxicity is not determined simply by the presence of toxic fragments, but also by other features in the compound which may modify the behaviour of toxicophores.

Figure 9.6 gives an example of the display of a DEREK answer for a query compound (part a) and display of the notes written by the rule writer for the rule which has been activated for this compound (part b). Prediction of the potential toxicity of compounds is of considerable appeal to most areas of the chemical industry, particularly if this reduces the need for animal testing. Unfortunately there is a major complication in the prediction of toxicity in the form of metabolism. Indeed, there are a number of therapeutic compounds which rely on metabolism to produce their active components. Currently, metabolic processes are dealt with in the DEREK system by explicit statements in the rule base [14] and for a method such as the identification of structural alerts, the alerting fragments may well have been chosen because human metabolism leads to toxic structures. In principle, the likely routes of metabolism for a given compound in a particular species can be predicted by an expert system, and programs exist which aim to do just that (see Section 9.4). The production of a generally applicable expert system for toxicity

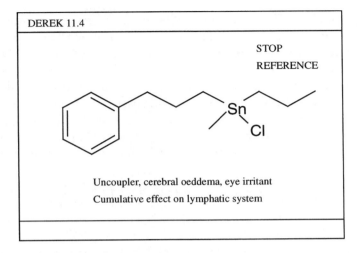

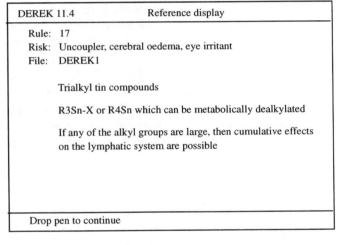

Figure 9.6 An example of the output from the DEREK program (reproduced from ref. [13] with permission of Wiley-Blackwell).

prediction is likely to require an expert system for metabolism predictions as well as perhaps some means of assessing distribution and elimination.

Other methods have been reported for the estimation of toxicity and these are often called 'expert systems' although the definition of an expert system used here, in which humans define the rules, would not label them as such. The TOPKAT program (Toxicity Prediction by Komputer Assisted Technology) uses a combination of substructural descriptors, topological indices, and calculated physicochemical properties to make predictions of toxicity [15]. The program has been trained on databases of a number of different toxicity endpoints using multiple linear regression if the experimental data is continuous (e.g. LD_{50}) or discriminant analysis if the data is classified. Prediction of toxicity for an unknown compound is made by calculation of substructural descriptors and physicochemical properties and insertion of these values into the regression or discriminant functions. One TOPKAT model has been created which links rat toxicity to mouse toxicity [16]. A 'rudimentary' model involving just these two activities is shown in Equation (9.1).

$$\log\left(\frac{1}{C}\right)_{\text{RAT}} = 0.636 + 1.694 \log\left(\log\left(\frac{1}{C}\right)_{\text{MOUSE}} + 1\right)$$
$$N = 160 \quad R^2 = 0.399 \quad S.E. = 0.41 \tag{9.1}$$

This equation uses a somewhat unusual double logarithm of the mouse toxicity data, to make this variable conform better to a normal distribution. A better predictive equation for rat toxicity was obtained by combining structural descriptors with the mouse toxicity data as shown in Table 9.4.

Input of chemical structure to the TOPKAT program is by means of SMILES; the program recognizes fragments and properties which correspond to the regression equation or discriminant function which is associated with the selected module (particular toxic endpoint) and calculates a value for the endpoint. The parts of the structure which

Table 9.4 Regression model for the prediction of rat LD_{50} from mouse LD_{50} and structural descriptions (reproduced from ref. [16] with permission of Sage Publications, Ltd).

WLN Key #	Variable description	Coefficient	F
	$\log\left(\log\left(\frac{1}{C}\right)_{\text{MOUSE}} + 1\right)$	1.654	227.3
144	Two heterocyclic rings	−0.361	12.4
	molecular weight	0.000432	11.9
10	One sulphur atom	0.125	11.4
39	One − C = O (chain fragment)	−0.0965	11.1
58	One −NH$_2$ group	−0.0990	7.76
	$(\log P)^2$	−0.00833	5.71
107	One heteroatom in more than one ring	0.191	7.00
37	One − OH group (chain fragment)	−0.0605	5.25
135	Two ring systems (not benzene)	0.186	5.15
67	One − C = O group (substituent)	−0.0856	4.56
	Constant	0.628	

$R^2 = 0.793 \quad SE = 0.33 \quad N = 160$

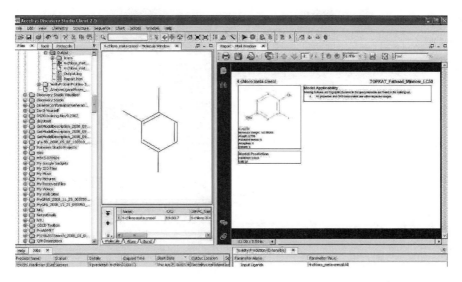

Figure 9.7 Computer screens from the TOPKAT program running through Discovery Studio showing a prediction of toxicity for 4-chloro-*m*-cresol against Fathead Minnow (figure kindly supplied by Accelrys).

are used in the calculation are shown graphically, as can be seen in the right-hand side of Figure 9.7.

The report in the figure gives details of the calculation of toxicity (LC_{50} for Fathead Minnow) for the compound, which in this case gives a value of 2.3 mg/l for LC_{50} (experimental value 7.4 mg/l, so not bad agreement). The program also recognizes features in a molecule that are not present in the predictive model, and will issue a warning if the compound has not been well 'covered'.

Figure 9.8 shows an example of a workflow that can be built for toxicity prediction from a file of structures (an SD file). In this case only one toxicity prediction module is used in the process but others can be easily added to give alternative results or a consensus prediction. The workflow was constructed using Pipeline Pilot (Accelrys, Inc.) which is a software environment which allows the assemblage of data analysis/handling routines using a graphical interface. The lower part of the figure shows a report for the prediction of toxicity of a herbicide to *Daphnia magna*. The report also gives details of the calculation and would warn if any of the molecular features were not recognized (covered) or if any properties were outside their normal range. The recognition of incomplete coverage is particularly important for any method that relies on the use

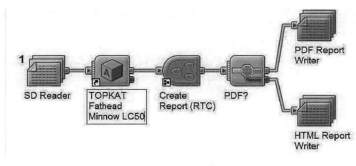

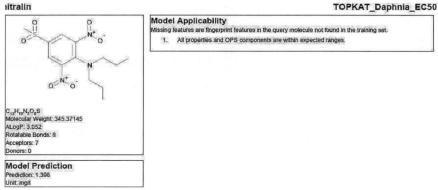

Figure 9.8 Illustration of a workflow system for toxicity prediction from a single module (TOPKAT), other prediction systems can be easily combined to form a consensus prediction. The lower part of the figure shows a prediction of toxicity for nitralin against *Daphnia magna* (figures kindly supplied by Accelrys).

of certain substructures or physicochemical features for prediction. It is essential to know how much of the molecule was recognized and used in the calculation in order to be able to make some sort of judgement of the reliability of prediction.

The CASETOX program uses a somewhat similar approach to TOPKAT but rather than using predetermined structural fragments, CASETOX generates all possible fragments for the compounds in a database [17]. Like TOPKAT, the CASETOX system has been trained on databases for a number of toxic endpoints. Perhaps one of the biggest problems with any of these toxicity prediction systems lies in the quality and suitability of the available databases. Since these databases often collect together results from different laboratories, the question of consistency obviously arises. Suitability is probably an even more important matter. If the compound for which prediction is required is a potential

Table 9.5 Results of carcinogenicity prediction (reproduced from ref. [18] with permission of the American Chemical Society).

Program or researcher	Equivocal compounds equated to non-carcinogen (%)	Equivocal compounds eliminated (%)
Programs		
MULTICASE	49	55
DEREK	59	62
TOPKAT	58	58
COMPACT	56	62
Humans		
Rash	71	68
Bakale	64	65
Benigni	63	72
Tennant & colleagues	75	84

pharmaceutical, then the best database for prediction will contain mostly pharmaceuticals, preferably with some degree of structural similarity. Such databases are not yet generally available.

Finally, how well do such prediction systems work? In 1990, a challenge was issued to interested parties to make predictions for 44 compounds that were then being tested for rodent carcinogenicity in bioassays by the National Toxicology Program (NTP) in America. The results of the predictions were discussed at a workshop in 1993 and are shown in Table 9.5 [18].[3]

The four computer programs, MULTICASE, DEREK, TOPKAT, and COMPACT, were not particularly impressive in their predictions since their best result was 62 % correct and the requested prediction was YES/NO, which might be expected to be 50 % right by chance, given an even distribution of carcinogens in the set. The results from the human groups or individuals are at least as good or considerably better, but in fairness it should be pointed out that in some cases these predictions made use of more information than was available to the programs. One other consideration should be borne in mind when comparing prediction results between computer systems, or between computers and humans, and that is the overall number of predictions made. The human experts generally made a prediction for every compound, whereas different computer systems omitted different numbers of compounds because their 'rules' could not cope with certain structures or sub-structures.

---

[3] At the time of the meeting, results had been obtained for 40 of the 44 compounds.

Benigni has reviewed the prediction of mutagenicity and carcinogenicity [19].

9.2.3   Reaction Routes and Chemical Structure

The LHASA system (Logic and Heuristics Applied to Synthetic Analysis) has already been mentioned as the basis of the DEREK toxicity prediction program. The LHASA program was originated as OCSS (Organic Chemical Simulation of Synthesis) by E.J. Corey [20] and is being further developed in the LHASA group of the chemistry department at Harvard University. This program contains a large database of organic reactions, also known as a knowledge base, and a set of rules (heuristics) that dictate how the reactivity of particular functional groups or fragments (retrons) is affected by other parts of the molecule, reaction conditions, etc. The aim of the program is to suggest possible synthetic routes to a given target molecule from a particular set of starting materials. This system actually starts at the target structure and breaks this down into simpler materials until eventually reaching the starting compounds, thus the process is retrosynthetic. An example of one of the steps given by the program for the synthesis of a prostaglandin precursor is shown in Figure 9.9.

When setting up the synthesis query, the user can select one of five different synthetic strategies, including a stereochemical option, and there is also an option for the program to make its own suggestions for strategies and tactics. Synthesis of a relatively complex molecule may proceed from simple starting materials by a great many different routes; Corey, for example, shows a retrosynthetic analysis of aphidicolin produced by LHASA which contained over 300 suggested intermediates [21]. Figure 9.10 shows an early stage in the synthesis planning of the prostaglandin precursor shown in Figure 9.9. Each number (node) on the display represents a different compound, Figure 9.9, for example, involved nodes 1 and 2.

Of course synthesis-planning expert systems can operate in the opposite direction, from starting materials to products, and the CAMEO program (Computer Assisted Mechanistic Evaluation of Organic reactions) is an example of this [22]. A problem with the operation of a synthesis-planning system in the forward direction, particularly if several steps are required, is the potentially large number of synthetic alternatives that must be considered.[4] Each intermediate molecule may undergo a number

---

[4] The same is also true of retrosynthetic systems.

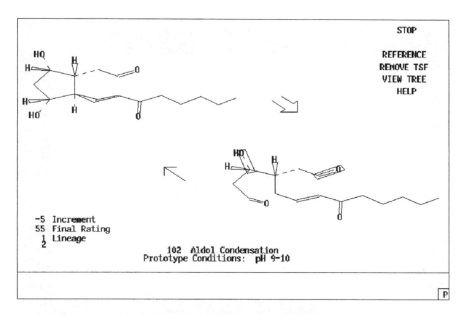

Figure 9.9 Computer screen from the LHASA program showing a reaction step (copyright LHASA).

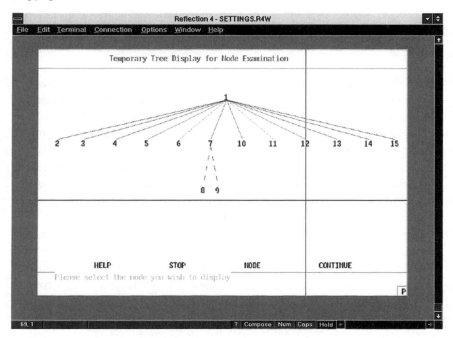

Figure 9.10 Computer screen from the LHASA program showing an intermediate stage in the breakdown of the prostaglandin precursor shown in Figure 9.9 (copyright LHASA).

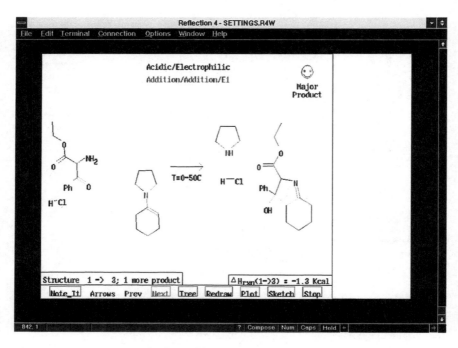

Figure 9.11 Output screen from the reaction prediction program CAMEO (copyright LHASA).

of transformations and thus the target may be reached by a number of different routes. A successful expert system not only has to work out the feasible routes, but also has to assign some likelihood of success, or degree of difficulty, to the individual routes. Figure 9.11 shows an example of an output screen from the CAMEO program for a reaction predicted using the acidic/electrophilic mechanistic module. This screen shows that this compound is one of the possible products, the smiling face symbol indicates that the program predicts that this will be a major product, and there is even a calculated ΔH for the reaction.

The CAMEO program has a quite comprehensive set of options for controlling the conditions under which reaction predictions will be made (Figure 9.12). The user can choose one of eight different mechanistic modules, temperature ranges may be set, and there are menus for the choice of reagents and solvents. Another view of the CAMEO system is that it is a reaction evaluation program. It may be used to verify suggestions, made by other programs or methods (or even individual chemists) concerning particular reactions within a sequence.

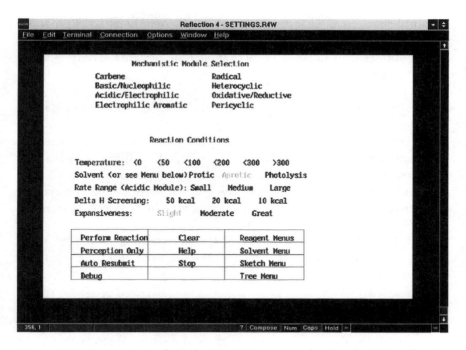

Figure 9.12 Output screen from the CAMEO program showing the choice of mechanistic modules and reaction conditions for prediction (copyright LHASA).

This account of the LHASA and CAMEO programs has been necessarily very limited; readers interested in further details should contact the suppliers of the software. Another approach to the problem of synthesis planning is to provide literature references to model reactions, thus allowing the chemist user to make his or her own assessment of feasibility, using his or her own expert system! A number of reaction database systems are in common use (e.g. REACCS, SYNLIB) and a combination of synthesis-planning software with literature reaction retrieval software is a powerful tool for synthetic chemistry.

The prediction of three-dimensional chemical structure from a list of atoms in a molecule and their connectivity is a good example of a chemical problem that may be solved by an expert system. We have already seen (Figure 9.2) how the SMILES interpreter can construct a two-dimensional representation of a structure from its one-dimensional representation as a SMILES string. The CONCORD program (CONnection table to CoORDinates) takes a SMILES string and, very rapidly, produces a three-dimensional model of an input molecule. This system is a hybrid between an expert system and a molecular mechanics program, molecular mechanics being the method by which molecular structures are 'minimized' in most molecular modelling systems. The procedure operates as follows.

1. The SMILES input string is checked for syntax errors and is made 'unique' if it is not already (see panel).
2. The atom symbols in the SMILES string are numbered sequentially and a connection table is constructed which indicates which atoms are bonded to each other.
3. A bond type table (i.e. single, double, aromatic, etc.) is constructed from the connection table and a ring table is constructed from the connection table.
4. The connection table and bond table are used to assign the number of hydrogens to be attached to each atom and other chemical validity checks are carried out.

A set of rules is used to assign reasonable three-dimensional structures to various features in the molecule, e.g. rings, and to certain bond angles and torsion angles. At this point the three-dimensional structure will be a good approximation for most parts of the molecule but some bond angles and torsion angles could be adjusted to minimize unfavourable steric interactions (at the expense of introducing extra energy into the system in the form of bond angle or torsion angle strain). This requires optimization of an expression for energy such as that shown in Equation (9.2).

$$E = \sum_{i=1}^{N} k_i^{l} (b_{0,i} - b_i) + \sum_{i=1}^{N} k_i^{\theta} (\Theta_{0,i} - \Theta_i) + \ldots \ldots \quad (9.2)$$

In this very simple expression for the energy of a molecule, the first two terms represent a summation over all the bonds of the compound of the energy contribution due to bond lengths and bond angles. The two constants, k_i^{l} and k_i^{Θ}, represent the force necessary to distort that particular type of bond (i) from its 'natural', i.e. average, bond length ($b_{0,i}$) or bond angle ($\Theta_{0,i}$). The standard bond lengths and bond angles have been derived from experimental measurements such as X–ray crystallography, and the force constants from spectroscopic data. Equation (9.2) is known as a molecular mechanics force field and can be elaborated to include contributions from torsion angles, steric interactions, charge interactions, and so on. The CONCORD system gives warnings of close

Table 9.6 Performance of CONCORD on a subset[a] of the Cambridge Structural Database (X-ray structures) and the fine chemical directory (FCD) and modern drug data report (MDDR)[b]. (Personal communication from R.S. Pearlman, College of Pharmacy, the University of Texas at Austin.)

	CSD	FCD and MDDR
Number of compounds	30,712	101,776
Total errors	3,207 (10.4%)	1,531 (1.5%)
made up from		
ring-system	707 (2.3%)	223 (0.2%)
chiral fusions	1,462 (3.4%)	216 (0.2%)
close-contacts	1,038 (3.4%)	1,092 (1.1%)
Close-contact warnings[c]	11,126 (36.2%)	28,080 (27.6%)

[a]Excluding 'unusual' atoms (e.g., As, Se, etc.).
[b]Compounds are often chosen for X-ray structure determination because they represent unusual structural classes, hence the relatively high error rate compared with the more 'typical' structures in the FCD and MDDR.
[c]Structures are generated in these cases and may be 'cleaned up' by molecular mechanics or quantum mechanics calculations.

interactions and other recognized problems in the final structure and these, of course, can be resolved manually or by structure optimization using another molecular mechanics program or by quantum mechanics. The combination of expert system and 'pseudo-molecular mechanics' generally does a good job, as shown in Table 9.6, and has the distinct advantage that it is very fast in terms of computer time. The input of SMILES strings to CONCORD can be automated and in this way many large corporate databases of (generally) good three-dimensional structures have been generated.

9.3 NEURAL NETWORKS

As was briefly mentioned in the introduction to this chapter, artificial neural networks (ANN) are attempts to mimic biological intelligence systems (brains) by copying some of the structure and functions of the components of these systems. The human brain is constructed of a very large number ($\sim 10^{11}$) of relatively slow and simple processing elements called neurons. The response time of a neuron (i.e. the time between successive signals) is of the order of a tenth to one-hundredth of a second. In computing terms this is equivalent to a 'clock speed' of 0.01 to 0.1 kHz, very slow compared with the processor speeds of commonly used personal computers (2 to 3 GHz). So what is it that makes man so smart? The answer lies in the fact that the brain contains a large number of processing elements which are working all the time; this is parallel processing on a grand scale, and the brain in computing terms is a massively parallel device. The other important feature of biological intelligence is the highly complex 'wiring' which joins the neurons together; a single neuron may be connected to as many as 100 000 other neurons.

So what do these processing elements do? Even a cursory examination of a textbook of neurobiology will show the complexity of the biochemical processes which take place in the brain. Various compounds (neurotransmitters) are involved in the passage of signals between neurons, and the functions of the neurons themselves are regulated by a variety of control processes. Ignoring the complexity of these systems the functions of a neuron can be summarized as follows.

1. the receipt of signals from neurons connected to it; these signals can be excitatory or inhibitory;
2. summation of the input signals, and processing of the sum to reach a 'firing' threshold;
3. the production of an output signal (firing) as dictated by (2) and transmission of this signal to other connected neurons.

This highly simplified description of how a biological neuron functions may not be a good model for the real thing but it serves as the basis for the construction of ANN. Intelligence in living biological systems appears to reside in the way that neurons are connected together and the 'strength' of these connections. Indeed, the creation of connections and the modification of connection weights is thought to be part of the processes involved in our development, i.e. learning and memory. It may not be clear where in the brain signals arise and which pathways they follow, although for certain regions of the brain, such as the sensory organs, it is more obvious. The eyes, for example, produce nervous signals in response to light and these are passed to the visual cortex. Some preprocessing of the information received by the eyes is carried out by sets of neurons which are organized in particular structures, e.g. layers. It is these three functions of biological neurons and their physical organization and connectivity which forms the basis of the construction of ANN.

ANN, like their biological counterparts, are built up from basic processing units as shown in Figure 9.13.

This artificial neuron receives one or more input signals, applies some kind of transformation function to the summed signal and produces an output signal (equal to the transformation) to be passed on to other

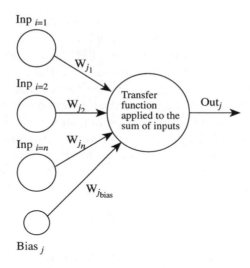

Figure 9.13 Diagram of an artificial neuron connected to three input neurons and a bias unit (reproduced from ref. [23] with permission of Springer).

neurons. A network usually receives one or more input signals and the input neurons, one for each input signal, behave somewhat differently in that they usually do not do any processing but simply act as distributors to deliver the signal to other neurons in the network. There are many ways in which the neurons in an ANN can be connected together, often referred to as the ANN 'architecture', but one of the most common is in the form of layers as shown in Figure 9.14.

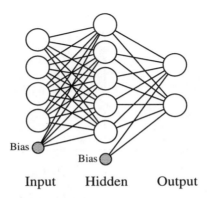

Figure 9.14 Illustration of how an ANN is built from layers of artificial neurons (reproduced from ref. [23] with permission of Springer).

This is known as a 'feed-forward' network since information enters the network at the input layer and is fed forward through the hidden layer(s) until it reaches the output layer. Each neuron in the input layer is connected to every neuron in the hidden layer and each hidden layer neuron in turn is connected to every neuron in the next layer (hidden or output). The strengths or weights of the connections between each pair of neurons are adjustable and it is the adjustment of these connection weights that constitutes training of a network. The neurons, or processing elements, in the hidden layer(s) and output layer apply a non-linear function (transfer function) to their summed inputs such as that shown in Equation (9.3),

$$OUTPUT = {}^{1}\!/\!\left[1 + e^{-s}\right] \tag{9.3}$$

where s represents the sum of the inputs to the neuron and Output represents its output signal. The shape of this function is sigmoid as shown in Figure 9.15 and thus the neuron mimics, to some extent, the way that biological neurons 'fire' when their input signals exceed some threshold. The use of a function such as that shown in Equation (9.3), or some other non-linear function, allows a network to 'build' non-linear relationships between its inputs and some desired target output.

The bias neurons, one for each layer, represent neurons which produce a constant signal. Their function is to act as shift operators so that the summed inputs for the neurons in the next layer are 'moved', on their transfer function scales, so as to produce signals. Training (adjustment of connection weights) is usually carried out in order to produce a desired

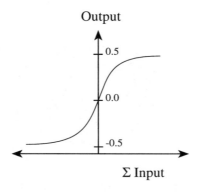

Figure 9.15 Representation of a commonly used transfer function. Scaled between −0.5 and +0.5, as often used in neural network programs. (reproduced from ref. [23] with permission of Springer).

target signal or signals at the output layer. A commonly used form of training is known as 'back-propagation of errors' and thus networks such as that shown in Figure 9.14 sometimes glory under the title of 'back-propagation feed-forward networks'. There are various ways in which connection weights can be assigned (before training) and networks can be trained. One very common procedure is to assign random values to the connection weights and then to repeatedly pass the training set data through the network, adjusting the weights by back-propagation, until some target error (Δ target–output) is achieved. Of course this does not guarantee that the 'best' solution has been reached and so the network is 'shaken' by applying small perturbations to the weights and then retraining, usually for a set number of passes through the data. This process can be repeated several times, storing the network connection weights and errors at the end of each training cycle, so that the 'best' network can be selected. For a more detailed description of the operation of neural networks see references [24] and [25] and the references contained therein. Reviews on the use of neural networks in chemistry have been published by Zupan and Gasteiger [26], Burns and Whiteside [27], Jakus [3], and Zupan and Gasteiger [28], and on the use of networks in drug design by Manallack and Livingstone [25] and Livingstone and Salt [29]. Applications of neural networks to a variety of problems in biology and chemistry are discussed in a volume of 'Methods in Molecular Biology' [30] and no doubt they have found uses in many other areas of science.

Finally, before moving on to the applications of ANN in data analysis, it is necessary to consider how networks are implemented. ANN are well suited to construction using dedicated computer hardware, particularly when we consider that they are meant to mimic a parallel-computing device. Hardware implementations have the advantage that they can be trained very quickly even when using very large data sets. The disadvantage of constructing networks in hardware, however, is that it is difficult or impossible to change the architecture of the network. Software implementations, although slower to train, are more versatile and are available, both commercially and as 'public-domain software', for a variety of computers.

9.3.1 Data Display Using ANN

One method of data display using artificial neural networks, the self-organizing map (SOM) or Kohonen map, has already been described in

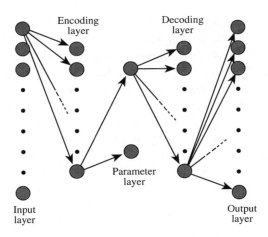

Figure 9.16   Diagram of a ReNDeR dimension reduction network (reproduced from ref. [31] with permission of Elsevier).

Section 4.3.2 of Chapter 4. As discussed in that section, the architecture of a SOM is quite different to the back-propagation feed-forward networks which are used in the majority of applications of ANN in data analysis. This section describes another, quite different, technique for displaying data using ANN.

The physicochemical properties that describe a set of molecules may be used as the input to a neural network and the training target may be some classification (discriminant analysis) or continuous dependent variable (regression analysis) as described in Section 9.3.2. The training target may also be the values of the input variables themselves and this is the way that a ReNDeR (Reversible Non-linear Dimension Reduction) network operates. A ReNDeR network (Figure 9.16) consists of an input layer, with one neuron for each descriptor, a smaller hidden layer (encoding), a parameter layer of two or three neurons, another hidden layer (decoding), and an output layer [31].

The encoding and decoding hidden layers are of the same size and there are as many output neurons as there are input. Each compound (or sample or object) in a data set is presented to the network by feeding in the values of its descriptor variables to the input neurons. The signals from the output neurons are compared to their targets, in this case the value of the input variables, for each compound and the weights in the network are adjusted until the output matches the input. Once training is complete the network has mapped the input onto the output by going through a 'bottleneck' of two (or three) neurons. Each sample in the data set can now be presented to the ReNDeR network in turn and a

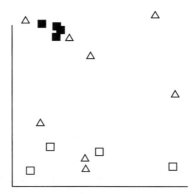

Figure 9.17   ReNDeR plot of a set of active (■), intermediate (Δ), and inactive (□) compounds described by 23 properties (reproduced from ref. [31] with permission of Elsevier).

value will be produced at each of the neurons in the parameter layer. These values will be the summation of the inputs, multiplied by their connection weights, received by the parameter neurons from the neurons in the encoding layer. The parameter layer numbers may be used as x and y (or x, y, and z) coordinates to produce a plot of the samples in the data set. Such a plot will be non-linear because of the network connections and non-linear transfer functions, and provides an alternative method for the low-dimensional display of a high-dimensional data set. An example of this type of display is shown in Figure 9.17 for a set of 16 analogues of antimycin-A_1 described by 23 calculated physicochemical parameters (these are the same compounds reported in Table 1.1).

The active compounds, shown as filled squares, are grouped quite tightly together and thus the plot might be expected to identify new compounds which will be active. The inactive compounds, along with three intermediates, lie in a quite different region of space in this display. For comparison, a non-linear map of this data set is shown in Figure 9.18 where once again it can be seen that the active compounds are grouped together, although there are some inactives nearby. The non-linear map in this case has made a better job of grouping compounds together which have intermediate activity.

Can we say that one of these two plots is best? The answer to that depends on the use that is to be made of the display, in other words what questions are we asking of the data? The ReNDeR plot gives a very clear separation between actives and inactives whereas the non-linear map groups most of the intermediates together. An encouraging thing is that the plots produced by the two different display methods are

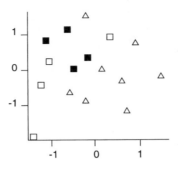

Figure 9.18   Non-linear map of the same set of compounds as shown in Figure 9.17 (reproduced from ref. [32] with permission of the American Chemical Society).

giving similar information about the data. Of course, we might expect that this non-linear display technique would give similar results to another non-linear method such as non-linear mapping. How do ReNDeR plots compare with linear displays such as those produced by principal components analysis (PCA)? Figure 4.8 shows a principal components scores plot for a set of analogues of γ–aminobutyric acid (GABA). These compounds were tested for agonist activity at the central nervous system GABA receptor, and the PC plot roughly separates them into potent and weak agonists and compounds with no agonist activity. The compounds were characterized by 33 calculated physicochemical properties and it was found that the scores plot could be considerably improved, in terms of its ability to classify the compounds, by selecting properties and re-computing the PCA [33]. A ReNDeR plot of this data is shown in Figure 9.19 where it can be seen that the compounds are quite clearly grouped according to their class of activity. This is an interesting result in that this technique is giving a superior result to the PCA display of the data. Since the network method is non-linear, this may show that the linear structure imposed by PCA is not suitable for this data set, although a non-linear map of the same data also failed to classify the compounds. The better classification by ReNDeR could, of course, be entirely fortuitous and it will be necessary to examine many other data sets to establish the utility of this new technique.

9.3.2   Data Analysis Using ANN

A neural network may be trained to reproduce any given target from a set of input values, provided it has a sufficient (see later) number of

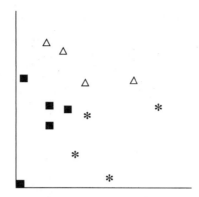

Figure 9.19 ReNDeR plot of GABA analogues (■ potent agonist, △ weak agonist, * no agonist activity) described by 33 properties (reproduced from ref. [31] with permission of Elsevier).

neurons and layers. Where the dependent variable or property to be predicted is classified, e.g. active/intermediate/inactive, the network can be set up with a neuron in the output layer corresponding to each of the classes. Training is carried out until only one neuron (the correct one!) is activated for each of the examples in the training set. This is equivalent to performing discriminant analysis (see Section 7.2.1), physicochemical descriptors are used as input to the network and, once trained, the network connection weights *might* be equated to the coefficients of the parameters in a discriminant function. Unfortunately, the connection weights cannot be identified quite so easily as this, since there will be a connection from every input neuron (parameter) to each of the neurons in the hidden layer. After training it may be found that most of these connection weights are around zero and one dominant weight may be found for a particular input parameter. However, it is just as likely that many of the connection weights will have 'significant' values and it will not be possible to extract the contributions made by individual variables (but see Section 9.3.4).

An example of a data set with a classified response which has been analysed using a neural network is shown in Table 9.7 [23].

This data set was chosen because the best discriminant function which could be generated from the sum of the parameters π, MR, and π^2 for the two substituents was only able to classify correctly 22 of the 27 compounds [34]. The network architecture used for this analysis consisted of two input units, one for $\Sigma\pi$ and one for ΣMR, a hidden layer and two output units, so that one unit could be activated (take

Table 9.7 Structure, physicochemical properties, and activity category of napthoquinones (reproduced from ref. [23] with permission of Springer).

Compound	R_1	R_2	$\Sigma\pi$	ΣMR	Category[a]
1	Cl	H	0.71	0.70	1
2[b]	OH	H	−0.67	0.38	1
3	OCH_3	H	−0.02	0.89	1
4	$OCOCH_3$	H	−0.64	1.35	1
5[b]	NH_2	H	−1.20	0.64	1
6	NHC_6H_5	H	1.37	3.10	1
7[b]	CH_3	CH_3	1.12	1.12	1
8	CH_3	OCH_3	0.54	1.35	1
9	OH	CH_3	−0.11	0.84	1
10[b]	Br	Br	1.72	1.78	1
11	Cl	$N(CH_3)_2$	0.89	2.16	1
12	OCH_3	OCH_3	−0.04	1.58	1
13	H	H	0.00	0.20	2
14	CH_3	H	0.56	0.66	2
15[b]	SCH_3	H	0.61	1.48	2
16	Cl	Cl	1.42	1.21	2
17	C_2H_5	H	1.02	1.13	2
18	$COCH_3$	H	−0.55	1.22	1
19[b]	SC_2H_5	H	1.07	1.94	2
20	OH	$CH_2C_6H_5$	1.34	3.28	1
21	OH	$COCH_3$	−1.22	1.4	1
22[b]	CH_3	SCH_3	1.17	1.94	1
23[b]	CH_3	SC_2H_5	1.63	2.4	1
24[b]	OH	Br	0.19	1.17	2
25	Cl	$NHCH_3$	0.24	1.63	1
26	OH	Cl	0.40	0.88	1
27	OH	NH_2	−1.87	0.82	2

[a]Category 1 = inactive; 2 = active.
[b]Compounds used for testing purposes in the second part of this analysis. Test compounds were chosen at random and the test set possesses approximately the same ratio of inactive to active compounds as in the original data set.

a positive value) for active compounds and the other unit for inactives. Network training was carried out using networks with different numbers of neurons in the hidden layer so as to assess the performance of the networks. It had already been shown that, given sufficient connections, ANN were able to make apparently successful prediction using random numbers [23]. This behaviour of networks is dependent on the number of network connections, the greater the number of connections the more easily (or more completely) a network will train. In fact it was pointed out by Andrea and Kalayeh [35] that the important quantity is not the overall number of connections but the ratio of the number of data

Table 9.8 Summary of network performance using 27 training compounds (reproduced from ref. [23] with permission of Springer).

Network architecture	Connections	ρ	Total RMS error	Summary (incorrect compounds)
2, 1, 2	7	3.86	0.4175	1, 7, 15, 17, 19, 24, 27
2, 2, 2	12	2.25	0.3599	1, 7, 8, 27
2, 3, 2,	17	1.59	0.2778	1, 7, 8
2, 4, 2	22	1.23	0.2750	1, 7, 8
2, 5, 2	27	1.00	0.2275	17, 24
2, 6, 2	32	0.84	0.1926	8
2, 7, 2	37	0.73	0.0309	All correct

points (samples) to connections which they characterized by the parameter, ρ.

$$\rho = \frac{\text{number of data points}}{\text{number of connections}} \qquad (9.4)$$

Table 9.8 shows the effect of adding extra hidden layer neurons to these networks; for a ρ value of 1.00 where there are as many connections as samples the network is able to classify all but two of the compounds successfully.

A better test of the suitability of ANN to perform discriminant analysis is to split the data set into separate test and training sets. This was done for this data to give a training set of 18 compounds and a test set of nine as indicated in Table 9.7. Results of training and test set performance are shown in Table 9.9 where it can be seen that training set predictions

Table 9.9 Summary of network performance and prediction using 18 training compounds (reproduced from ref. [23] with permission of Springer).

Network architecture	Con[*]	ρ	Total RMS error	(incorrect) compounds)	Prediction summary (incorrect compounds)
			Training summary		
2, 1, 2	7	2.57	0.3166	1, 27	7, 10, 15, 19, 24
2, 2, 2	12	1.50	0.2059	1	2, 5, 7, 10, 15, 19, 24
2, 3, 2	17	1.06	0.3099	1, 27	7, 10, 15, 19, 24
2, 4, 2	22	0.82	0.2359	27	7, 10, 15, 19, 23, 24
2, 5, 2	27	0.67	0.2357	27	7, 10, 15, 19, 23, 24
2, 6, 2	32	0.56	0.0274	All	7, 10, 15, 19, 24
2, 7, 2	37	0.49	0.0367	All correct	2, 5, 7, 10, 15, 19, 24

[*]The number of connections in the network.

are good even for the highest value of ρ and at the lower ρ values all compounds are predicted correctly.

Prediction performance for the test set, on the other hand, was uniformly bad and if anything got even worse at the lowest values of ρ. This demonstrates that the ANN is able to fit the data better than a linear discriminant function, probably because of the non-linearity[5] involved in the ANN modelling, but that the fitted model is not very useful in prediction. This may be because the properties used to describe the molecules are insufficient to characterize their behaviour sufficiently well, or it may be that the neural networks have been 'over-trained' as discussed in the next section.

An example of the use of neural networks to classify olive oil samples described by pyrolysis mass spectrometry data shows that ANN can work well in prediction [36]. In this work a training set of extra-virgin olive oils and adulterated oil samples (added peanut, sunflower, corn, soya, or sansa olive oils) were analysed by pyrolysis mass spectrometry to give spectra in the M/Z range of 51–200. Cluster analysis and canonical variates analysis of these data showed that the oil samples were broadly classified on the basis of the cultivar from which the extra-virgin oil was derived; extra-virgin and adulterated samples were not distinguished. A three layer back-propagation network with 150 input neurons (one for each M/Z value), eight hidden neurons, and one output neuron was trained with the training set data and found to predict all of the training samples successfully. This is perhaps not surprising since the training set was very small (24 samples) compared with the number of connections in the network (1217)[6]. Network performance on an unknown test set (samples were analysed blind) was very good, however, as shown in Table 9.10.

ANN may be used to fit a continuous response variable to a set of physicochemical properties, the network just requires one output unit and the training targets are the values of the response variable (IC_{50}, ED_{50}, etc.) for each compound in the set. Performance of these networks, however, can be deceptively good if care is not taken with the network architecture [37]. Figure 9.20 shows the results of network training using random numbers in which four columns of random numbers were used as input data and a column of random numbers was used as the target

---

[5] The discriminant function also employed non-linearity by using a π^2 term, but this may not be the appropriate function to use for successful modelling of this data set.
[6] The network has 150×8 weights between input and hidden layer, 8×1 between hidden and output, and 8×1 plus 1×1 for the bias units.

Table 9.10 Network prediction of test set oil samples (reproduced from ref. [36] with permission of Wiley-Blackwell).

Codename	Network answer[a]	Virgin or adulterated
Perugia	1	Virgin
Lecce	1	Virgin
Urbino	1	Virgin
Rimini	0	Adulterated
Taormina	0	Adulterated
Napoli	1	Virgin
Milano	1	Virgin
Trieste	1	Virgin
Torino	0	Adulterated
Cagliari	0.8[b]	Virgin
Bolzamo	1	Virgin
Venezia	0	Adulterated
Roma	0	Adulterated
Genova	1	Virgin
Bari	1	Virgin
Pescara	0	Adulterated
Padova	0	Adulterated
Palermo	0	Adulterated
Firenze	1	Virgin
Ancona	1	Virgin
Siena	0	Adulterated
Messina	0	Adulterated
Bologna	0	Adulterated

[a]The network was trained and interrogated five times. The scores given are the average of the five runs (± 0.001), where virgin is coded 1 and adulterated oil is coded 0.
[b]The network indicated that the oil Cagliari was of virgin quality (1) on four of the five trainings.

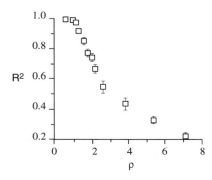

Figure 9.20 Plot of R^2 versus ρ for regression networks using random numbers (reproduced from ref. [37] with permission of the American Chemical Society).

(dependent variable). Any attempt at measuring predictive ability for these networks would be meaningless but the fit can be assessed, as a correlation coefficient (R^2) by comparison of the network output with the target values. As can be seen from the figure, quite high R^2 values are achieved below $\rho = 2$.

Finally, ANN may be used to build models for multiple dependent variables by simply including an output neuron for each of the dependent variables. Section 8.5 showed how two measured molecular properties, partition coefficient (log$P$) and solubility (log$S$), could be modelled simultaneously using PLS and canonical correlation analysis. This set consisted of a training set of 552 molecules and a test set of 68 all characterized by 37 calculated molecular descriptors [38]. The fit of multiple linear regression, canonical correlation and PLS models to this data is shown in Table 9.11.

ANN were constructed for the two dependent variables separately using the descriptors chosen by multiple linear regression (28 descriptors for log$S$ and 26 for log$P$) and a single ANN was computed with two output neurons to model log$S$ and log$P$ simultaneously. Network architecture and the optimal training endpoint (see Section 9.3.3) were determined on the results obtained for a further validation set of 68 compounds. The network for log$S$ used 6 neurons in the hidden layer and that for log$P$ had 5 hidden layer neurons. As can be seen from the

Table 9.11 Comparison of the modelling results for MLR, CCA, PLS and ANN on Log$S$ and Log$P$ data.

Model	Nvars	Training set		Test set	
		R^2	s	R^2	s
		Log$S$			
MLR	28	0.78	0.75	0.78	0.75
CCA	37	0.75		0.76	
PLS	37	0.63	0.98	0.61	0.99
ANN	28	0.87	0.60	0.83	0.65
ANN	37	0.89	0.53	0.83	0.65
		Log$P$			
MLR	26	0.87	0.65	0.86	0.65
CCA	37	0.85		0.87	
PLS	37	0.7	0.98	0.68	0.90
ANN	26	0.91	0.53	0.89	0.55
ANN	37	0.93	0.47	0.89	0.55

Nvars is the number of variables used in the model. The PLS models used 5 latent variables. The ANN models of fewer than 37 variables used the same variables as chosen by multiple linear regression.

table, MLR and CCA give roughly comparable results in terms of fit and prediction while the PLS models are poorer. The PLS modelling was stopped at 5 latent variables since, as can be seen in Figure 8.14, there is a discontinuity in a scree plot of RMSEP at 4 dimensions. The ANN, on the other hand, give better fits and predictions for both properties with the combined output network performing slightly better, although this did use all 37 independent variables rather than the subsets. Since the neural networks are using the same information as the other methods this better performance is presumably due to the ability of the ANN to include non-linearity in the models. There is a cost to this improved performance, of course, and that is in terms of ease of interpretation. The MLR models are relatively easy to interrogate since they 'just' consist of 28 or 26 regression coefficients for the selected variables. The canonical correlation equations are more involved since there are two pairs of them and thus two coefficients for each of the responses (loading onto the first and second canonical variate of the first set) and two coefficients for each of the independent variables. The PLS models are a degree more complex since they involve coefficients for 5 dimensions but the ANN models are the most obscure of all since the information in the model is contained in the connection weights between neurons. In the case of the combination network, which had a hidden layer of 10 neurons, this means a total of 402 connections since there are 37 inputs, plus a bias neuron, and 10 hidden neurons plus a bias. Such ANN models tend to be treated as a 'black box' with no attempt made at interpretation although there are ways in which these models may be probed as described in Section 9.3.4.

9.3.3   Building ANN Models

Many neural network software packages have built-in tools and guidance for the construction and training of ANN but it is important to be aware of the potential problems and pitfalls in their use. Quite a wide variety of different types of artificial neural network have been developed but the majority of applications in data analysis have involved the self-organizing map (SOM) and back-propagation feed-forward (BPN) networks. Construction of a SOM is fairly straightforward as it mostly consists of choice of the resolution (number of neurons) of the resulting plot. The number of input neurons is decided by the choice of input parameters and training is continued until a 'useful' picture emerges.

Successful development of a BPN is more involved and thus there are a number of steps to consider:

1. choice of network architecture and initialization of weights;
2. choice of transfer functions;
3. selection of a training algorithm to adjust the weights;
4. division of the data into training, test and possibly validation sets;
5. decision on when to stop training.

1. *The choice of network architecture* involves decisions on how many hidden layers to use, how many neurons in the hidden layers, and how many neurons to use in the output layer. In principle, there can be a number of hidden layers but in practice it has been found that one layer is usually sufficient. Choice of the number of output neurons is usually easy since only one is needed if training is to a single dependent variable. The first example shown in the previous section where the dependent variable was classified used two output neurons but the same effect could have been achieved by training a single output neuron to produce 0 or 1 for the different classes. The major problem in the determination of an appropriate network architecture is the number of hidden layer neurons. What is required is a sufficient number of hidden neurons so that the network can develop enough complexity to model the data but not too many because this can cause the network to be slow to train and, more importantly, give rise to overfitting. As we saw in the previous section it is essential to have more data points than connections in the network ($\rho > 1$ in Equation (9.4)) otherwise the network will be able to model the data perfectly, by effectively training 1 connection per sample, but will be unable to make predictions. Initialization of the weights is normally carried out by random assignment using values in a suitable range, where the range of weights will be determined by the training algorithm employed. This helps to ensure that training will lead to a useful network but it also means that retraining a network from scratch will often yield a different network. This is a feature of ANN; there is no unique solution to a BPN unlike other data modelling techniques. This is discussed further at the end of this section.

2 & 3. There are a number of different *transfer functions* which can be used to introduce non-linearity into the ANN models and choice of these will to some extent be dictated by what is available in the particular software package used. Similarly there are several different *training algorithms* available to adjust the weights during training and these all have advantages and disadvantages in terms of speed to convergence, ability

to escape local minima and so on. Choice of these are also dictated by their implementation in the software package employed.

4 & 5. *Division of the data into training, test and possibly validation sets* and the *decision on when to stop training* lie at the heart of successful data modelling using ANN. There are no 'rules' here and the number of samples available dictates the size of any sets but the main aim is that any subsets should be representative of the overall data set. The strategy of using a training set to fit a model and a test set to judge its predictive performance has already been discussed, but for ANN the procedure can be somewhat different in that this division can be used as a means to stop training. The problem with training ANN is that any weight optimization algorithm will carry on adjusting the weights until as perfect a possible fit is obtained. This will usually be at the expense of predictive performance because it is possible to 'overtrain' a network so that it learns all the peculiarities of the training set data without fitting a generalizable model. The network will probably have fitted the 'correct' model as part of the training process but will then have carried on adjusting the weights to fit the training data better while at the same time moving away from the proper model. A way to avoid this is to use a process known as 'early stopping' or stop-training [39]. Figure 9.21 shows some data which illustrates this process.

In early stopping the training set data is used to provide the targets for the weight adjustment algorithm and then periodically, say at the end of every 10 cycles through the training data, another set of samples are presented to the network for prediction and the prediction error is calculated. In Figure 9.21 the training set error (labelled learning) is shown as a solid line and it can be seen that this continues to decrease with increasing training iterations. This is a common feature of the algorithms used to adjust the weights and training can continue until the change in training set error falls below some pre-set minimum. This is dangerous, though, since as can be seen from the figure the mean square error for a separate data set, labelled the control set, begins to rise at an earlier stage in the training. There are three stopping points (S_1, S_2 and S_3) indicated by arrows on the figure. The last one, S_3, is where training might ordinarily be stopped as this is where the training set error is almost constant but it can be seen that the errors for the validation set (labelled control) and the joint set of validation and training are increased at this point. Thus, early stopping involves the choice of S_1 or S_2 as a stopping point (there is little to choose between them). This example has highlighted the problems with naming the data sets used in ANN training. A set of data used to train the network is generally called a training set, although

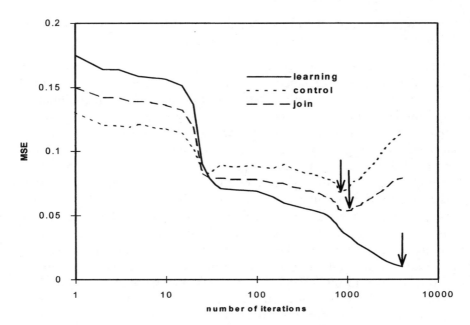

Figure 9.21   Plot of mean square error versus the number of training epochs (iterations) for a training set, validation set and joint set of data points for a linear artificial data set. The three arrows represent early stopping points S_1, S_2 and S_3 respectively (reproduced from ref. [39] copyright (1995) American Chemical Society).

here it was called learning, but the set used to judge an early stopping point is like a test set, but since the data it contains is used as part of the training process it isn't an independent test set in the normal use of the term. Thus, it is often called a validation set, although here it was called control. Finally, a true test set is a third set that has taken no part in the training process.

As was briefly mentioned earlier, a trained BPN is not a unique solution but simply a minimum, in terms of the error between the outputs and targets, from a particular starting set of network connection weights. Thus, a commonly employed technique is to train a set of neural networks and then take a subset of the 'best' networks, that is those with the lowest training errors, to form a committee in order to make predictions. This has the advantage that for a continuous response variable it is possible to assign an error range to the predictions and thus give some measure of uncertainty in the predicted values of any new samples. Such an approach may be variously called a 'consensus' network [40] a network ensemble [39] or a committee of neural networks [41]. An example of this is a study which aimed to identify molecules which

Table 9.12   Confusion matrices for consensus network predictions for kinase targets (reproduced from ref. [42] with permission of the American Chemical Society).

	Active	Inactive	Total	%Correct
	Training set predictions			
active	105	15	120	87.50
inactive	5	115	120	95.83
total	110	130	240	
%correct	95.45	88.46		91.67
	Validation set predictions			
active	45	15	60	75.00
inactive	4	56	60	93.33
total	49	71	120	
%correct	91.84	78.87		84.17
	Test set predictions			
active	44	16	60	73.33
inactive	9	51	60	85.00
total	53	67	120	
%correct	83.02	76.12		79.17

would be active against a particular gene family of protein targets [42]. In this work, compounds were described by a set of 20 calculated molecular properties, neural networks were constructed with a hidden layer of 3 neurons and an output layer of 1 unit and 1000 networks were trained from the usual randomly assigned starting weights. The 100 best networks were chosen to form a consensus and Table 9.12 shows the results for a set of kinase targets.

As can be seen from the table the data was split up into 3 sets; a training set of 240 compounds, a validation set of 120 and a test set of 120. The overall performance for the training set was about 92 % with the expected reduction in performance for the validation set and the unseen test set. The use of a full confusion matrix such as this to report the results shows that the networks performed better at classifying active compounds as opposed to inactives. Since the aim of the study was to classify compounds which were going to be purchased for screening against kinase targets then it is better to be more certain of the compounds which are likely to be active, and hence potentially useful.

Finally, it has already been shown that a genetic algorithm may be used to select multiple linear regression models from a large set of independent variables (Section 6.3.1.5). This is essentially an efficient method to find solutions in a very large potential solution space. The possibilities for ANN models, in terms of network architecture, choice of input variables, network connection weights and so on also represent solutions in a very large solution space and so genetic methods have been applied to solve these problems. There is something quite intellectually satisfying in making use of two methods borrowed from nature to solve problems of data analysis.

9.3.4   Interrogating ANN Models

One of the drawbacks in the use of ANN to model data is that the model is hidden within a large collection of network connection weights. This is fine if all that is required of a model is the ability to predict but if we want to try to interpret or understand a model then it isn't very satisfactory. There are, however, ways to interrogate a trained neural network. One technique involves presenting a constant signal at all of the input neurons except one, varying the signal to this one input neuron and examining the output. The aim of this is to see how a particular independent variable affects the dependent while being fed through the network model. One problem with doing this is the choice of signal to apply to the neurons which are kept constant. Should it be the mean value of each of these neurons or the value of the variable for one of the samples in the set, perhaps the sample with a median value of the dependent variable?

Another technique is known as sensitivity analysis. There are various flavours of this method, known in the neural network field as pruning since they generally aim to remove unnecessary connections between neurons, but they can be broadly classified as magnitude based or error based. Magnitude based techniques depend on the direct analysis of neuron weights while error based methods take account of changes in network error following the elimination of connections. Some of these techniques have rather marvellous names such as 'optimal brain surgeon' and 'optimal brain damage' which just reflects the biological origins of ANN. A simple example may serve to illustrate sensitivity analysis. A data set of charge-transfer complexes has already been introduced in Section 7.3.1. This set consists of experimental measurements of complex formation for simple monosubstituted benzenes described by 58 computed physicochemical descriptors, the independent variables [43]. A subset of 11 descriptors were selected on the basis of their correlation with the dependent variable, κ, derived from the experimental measurements. Neural networks with the architecture 11:5:1, that is to say 11 input neurons, 5 hidden neurons and a single output neuron, were set up and initialized with random starting weights. The networks also included a bias neuron on the input and hidden layers. Neural networks were trained 400 times and the sensitivities of the input parameters calculated [44] for each of the input properties. The least sensitive (least useful) variable found in the first run was variable 11 which agrees with the order in which these properties were originally chosen; they are ordered 1 to 11 in terms of their individual correlations with κ. This variable was removed and the network reconstructed (10:5:1) and again refitted 400 times. The next least sensitive variable found at this stage was variable 6, not variable 10 as might have been expected from the correlations with κ. Complete results from this sensitivity analysis are shown in Table 9.13.

The first row of the table shows the results for a network which contained all 11 properties, the next row shows that variable 11 (Sn(3)) was omitted, then variable 6 (Sn(1)), then variable 7 and so on. Elimination of variable 11 in the first step is the expected result since this variable had the lowest individual correlation with κ but the next step in the procedure shows that a variable with a higher correlation than four others (7,8,9 and 10) is eliminated. Variable 7 (Sn(2)) is the next to be dropped and then, surprisingly, the variable with the third highest correlation with κ (E_{HOMO}). The three 'best' variables chosen by the networks (CMR, ClogP and P3) give a 3 term multiple linear regression

Table 9.13 Network fitting and eliminated properties for the charge-transfer data set (reproduced from ref. [45] with permission of Springer).

1	2	3	4	5	6	7	8	9	10	11	q²(S1)
CMR	ClogP	E_{HOMO}	P3	Mux	Sn(1)	Sn(2)	P1	Fe(4)	Mu	Sn(3)	0.95 ± 0.01
										x	0.95 ± 0.009
					x					x	0.95 ± 0.009
					x	x				x	0.95 ± 0.008
		x			x	x				x	0.95 ± 0.008
		x			x	x	x			x	0.94 ± 0.009
		x			x	x	x		x	x	0.94 ± 0.007
		x			x	x	x	x	x	x	0.94 ± 0.006
		x		x	x	x	x	x	x	x	0.90 ± 0.007
		x	x	x	x	x	x	x	x	x	0.90 ± 0.006
	x	x	x	x	x	x	x	x	x	x	0.32 ± 0.03

Eliminated parameters are denoted by x. The correlation coefficients (q²) and their 95 % confidence limits were calculated by leave-one-out cross-validation.

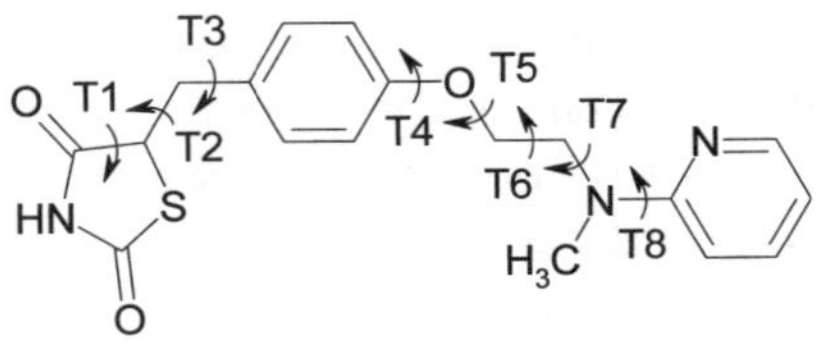

Figure 9.22 The molecular structure of rosiglitazone with the 8 flexible torsion angles (T1 to T8) used for conformational analysis (reproduced from ref. [50] with permission of Springer).

equation with an R^2 of 0.92, the best 3 term equation (found by forward inclusion linear regression) for CMR, ClogP and E_{HOMO} has an R^2 of 0.95. There is clearly some underlying non-linear relationship between κ and the physicochemical descriptors and it is this non-linearity which is presumably responsible[7] for the different order in which variables are selected by the network. A non-linear model involving the first few parameters chosen by the networks may give a better fit to the κ data than the multiple linear regression equations.

These approaches have all looked at the effect or contribution of individual descriptors but it is the combination of variables, along with non-linearity and cross terms introduced by the BPN, which goes to make up the complete model. Is it possible, therefore, to interrogate the entire model? The answer to this question is, perhaps surprisingly, yes. There are a number of different methods for extracting rules from trained networks such as NeuroRule [46], BioRe [47], MofN3 [48], REANN [49], TREPAN [50] and others. Many of these techniques generate a set of rules known as M-of-N rules. Rules in this form state, 'If M of the N conditions, $a_1, a_2, \ldots, a_m$ are true, then the conclusion b is true.' It has been argued [51] that some concepts can be better expressed in this form than the other logical 'if-then' form of rules and it seems that this representation also helps to avoid the combinatorial explosion in tree size found with if-then rules. An example of the use of TREPAN involves molecular dynamics simulations of the antidiabetic agent rosiglitazone shown in Figure 9.22.

The data involved are the dihedral angles of the 8 flexible torsion angles indicated on the diagram. A sample structure was taken every 1 picosecond during the course of a 5 nanosecond simulation, leading to 5000 data points describing the simulation. Each of these conformations

---

[7] Although it is possible that collinearity or multicollinearities amongst the descriptors may also contribute.

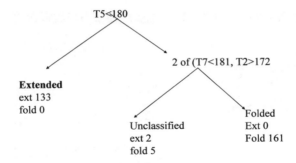

T5<180

2 of (T7<181, T2>172

Extended
ext 133
fold 0

Unclassified
ext 2
fold 5

Folded
Ext 0
Fold 161

Figure 9.23 The TREPAN M-of-N rules for classification of the conformations of rosiglitazone (reproduced from ref. [50] with permission of Springer).

was classified as either a folded or extended structure based on the distance between the two ends of the molecule; from the simulations carried out, the conformations were divided approximately 50:50 between these two states. Application of a datamining algorithm called C5 [52] resulted in highly accurate classifications of the conformations but using a very complex decision tree [53]. A neural network trained on these data gave a slightly lower accuracy of prediction but application of the TREPAN algorithm to the trained network resulted in the very simple set of rules shown in Figure 9.23.

This has been a rather superficial treatment of an important topic but detailed discussion of these methods is beyond the scope of this book. Interested readers should consult the cited papers and the references therein.

9.4   MISCELLANEOUS AI TECHNIQUES

The expert systems described in Section 9.2 should illustrate some of the principles of the construction and operation of expert systems in chemistry. Given a suitable knowledge base (empirical database) and set of production rules it is possible to predict various chemical properties from structure. Many such systems exist (for example, spectroscopic properties, solubility, heat of formation, etc.), although they are not always called 'expert systems'. The Rekker system for log P [4] has been coded into a computer-based expert system called PrologP. This system operates in a very similar way to the CLOGP calculation routine, taking a graphical input of structure, dissecting this into fragments, and then applying the rules of the Rekker scheme to calculate log P. The same

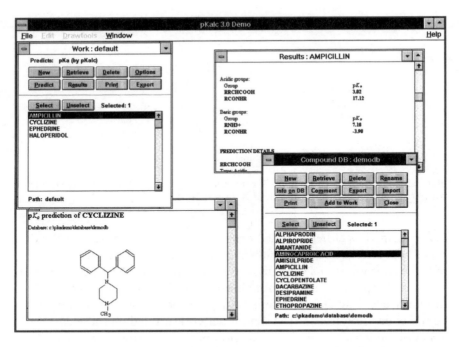

Figure 9.24 Computer screen from the program pKalc (reproduced with permission of Compudrug Company Ltd.).

company has also produced a pK_a prediction expert system (pKalc) which uses a Hammett equation (for aromatic systems) or a Taft equation (for aliphatics) as a basis for the calculation. Once again, the program takes graphical input of structure which is dissected into fragments. The ionizable groups are perceived and the appropriate equation selected for the prediction of the dissociation constant of each group. The rest of the molecule is treated as fragments or substituents which will modify the pK_a values and fragment constants, equivalent to σ values, are looked-up in a database and applied to the prediction equations. An example of an output screen from this program is shown in Figure 9.24; the program has the facility to sketch in molecules, as shown for cyclizine, store compounds in a database, and predict pK_a values, as shown for ampicillin. Ionization, of course, affects partition coefficients since it is generally the un-ionized species which partitions into an organic phase.[8] The PrologP and pKalc programs have been combined to create a

---

[8] Ionized species can dissociate into a 'wet' organic phase singly and as uncharged ion pairs.

distribution coefficient prediction system (PrologD) where log D represents the partitioning of all species at a given pH.

Before leaving expert systems, it is worth considering two problems involved in the prediction of toxicity, namely metabolism and distribution. These two problems are related in that metabolizing enzymes are variously distributed around the body. Thus, the distribution characteristics of a particular compound or its metabolites may dictate which elimination systems they will encounter. Similarly, as metabolism proceeds, the distribution properties of the metabolites may encourage them to migrate into different tissues. Prediction of toxicity necessarily involves the identification of at least the major products of metabolism and of course the situation is further complicated since different species will be eliminated at different rates. The DEREK system makes some attempt to account for metabolism by incorporating some well-known predictive rules. Other systems deal with specific metabolizing enzymes, for example, the COMPACT program deals with cytochrome P_{450} [54], while at least one program (METABOLEXPERT) attempts to combine metabolic pathways with a simulation of pharmacokinetic behaviour. While these systems have not yet reached the reliability of log P prediction programs, an inherently simpler problem, it seems inevitable that they will improve as the body of data is increased.

Rule induction is an artificial intelligence method that has been applied to the analysis of a number of chemical data sets. As the name implies, rule induction aims to extract rules from a set of data (descriptor variables) so as to classify the samples (compounds, objects) into two or more categories. The input to a rule induction algorithm is a number of test cases, a test set, and the output is a tree-structured series of rules, also known as a class probability tree. A popular rule induction algorithm is known as ID3 (Iterative Dichotomizer three) [55] and its operation in terms of information can be described as follows. If a test set contains p samples of class P and n samples of class N, a sample will belong to class P with probability $p/(p + n)$ and to class N with a probability $n/(p + n)$. The information in a decision tree is given by

$$I(p, n) = -p/(p + n)\log_2 p/(p + n) - n/(p + n)\log_2 n/(p + n) \qquad (9.5)$$

If a particular feature (property, descriptor), F, in the data set, with values $(F_i, F_{i+1}.....)$ is used to form the first rule of the decision tree, also known as the 'root' of the tree, it will partition the test set, C, into C_i, C_{i+1}, and so on, subsets. Each subset, C_i, contains those samples which have value F_i of the chosen feature F. If C_i contains p_i samples of class P and n_i samples of class N, the expected information for the subtree C_i is $I(P_i, n_i)$. The expected information required for the tree with feature F as a root is obtained as the weighted average

$$E(F) = \sum_{i=1}^{v} (p_i n_i)/(p + n) I (p_i, n_i) \qquad (9.6)$$

The information gain on branching on feature F is given by Equation (9.7).

$$gain(F) = I (p, n) - E (F) \qquad (9.7)$$

The ID3 procedure examines all the features in the data set and chooses the one that maximizes the gain, this process being repeated until some pre-set number of features are identified or a particular level of reliability is achieved. One problem with this procedure is that 'bushy' trees can be produced, that is to say decision trees which have so many rules that there is a rule for every one or two samples: the ID3 algorithm can be modified, using a significance test, to reject rules that are irrelevant [56].

Examples of the application of the ID3 algorithm to four sets of data involving biologically active compounds have been reported by A-Razzak and Glen [56]. One of these consisted of an expanded version (17 compounds) of the set of 13 γ–aminobutyric acid analogues (GABA) already shown in Figures 4.8 and 9.19. This particular set of compounds was described by seven computed physicochemical properties which did a very reasonable job of separating activity categories, as may be seen from the non-linear map shown in Figure 9.25.

The ID3 algorithm was run on a larger set of 24 computed properties to give the decision tree shown in Figure 9.26. Interpretation of this tree is fairly self-evident; the data set is split into two above and below a surface area value of 162.15, for example. This is one of the attractions of this form of 'machine learning', the decision rules may be readily understood and should be easy to apply when attempting to design new molecules. Two of the three properties used to provide these decision rules were included in the set of seven parameters used to produce the non-linear map. However, if one is interested in examining the effect of particular variables, perhaps because they have proved important in the prediction of another activity, the ID3 algorithm can be forced to create decision rules for user-selected features. In this example, the samples fell naturally into three classes; where the dependent variable is continuous,

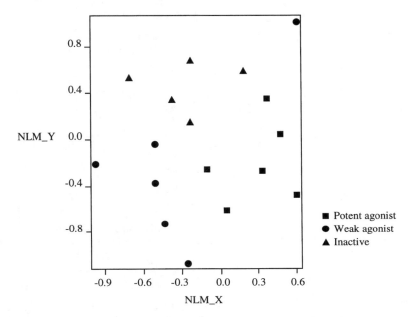

Figure 9.25 Non-linear map of 17 GABA analogues described by seven physico-chemical properties (reproduced from ref. [56] with permission of Springer).

the ID3 algorithm can be used by classifying compounds according to a range of the response variable.

Another example of the use of the ID3 algorithm is given in a report which compares 'rule-building expert systems' with pattern recognition

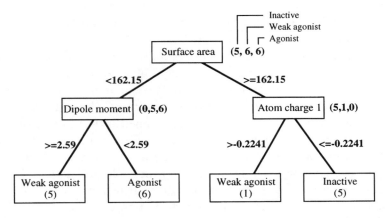

Figure 9.26 Decision tree from the ID3 algorithm run on the GABA analogues shown in Figure 9.25 (reproduced from ref. [56] with permission of Springer).

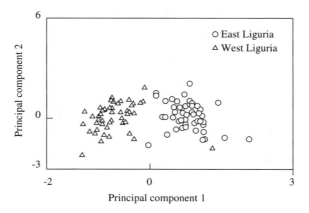

Figure 9.27   Principal components scores plot for olive oil samples characterized by their content of eight fatty acids (reproduced from ref. [57] with permission of the American Chemical Society).

in the classification of analytical data [57]. Figure 9.27 shows a principal components plot of 100 olive oil samples characterized by their content of eight fatty acids.

The samples are clearly separated into two geographical areas, Eastern and Western Liguria. An implementation of the ID3 algorithm called EX-TRAN was able to characterize these samples on the basis of their content of linolenic acid (first rule or root of the tree), oleic acid, linoleic acid, and palmitic acid. A comparison of the performance of EX-TRAN with k–nearest neighbours (one and five neighbours) and linear discriminant analysis (LDA) is shown in Table 9.14, where it can be seen that the results are slightly worse than the pattern recognition techniques (except KNN with raw data).

The final artificial intelligence method for the analysis of chemical data that will be discussed in this section might also be called 'rule-building expert systems'. The widespread use of molecular modelling packages in drug design has led to the creation of 'pharmacophore' or 'biophore' recognizing systems. A pharmacophore is defined as that pattern of atoms (or perhaps properties) which is required to exist in a molecule in order for it to exert some particular biological effect. It is generally accepted that the pharmacophore is 'recognized' by the biological activity site and presumably some, if not all, parts of the pharmacophore are involved in the compound binding to the site. A biophore has a less restrictive definition in that a biophore is some pattern of atoms and/or properties which occurs in some of the active or inactive molecules. The concept of a biophore for inactivity is an interesting one; presumably this relates to

Table 9.14 Comparison of validation results: number of wrong prediction results (not counting object 50 of eastern Liguria) (reproduced from ref. [57] with permission of the American Chemical Society).

	Eastern Liguria	Western Liguria
EX-TRAN	2	2
1NN		
Raw data	1	5
Autoscaled	1	2
Range-scaled	0	2
5NN		
Raw data	0	5
Autoscaled	1	1
Range-scaled	1	0
LDA		
Stepwise	0	0
All variables	0	1

a pattern of atoms and/or properties which are responsible for confusing recognition at the active site, or perhaps preventing binding by some repulsive interactions.

The CASE (Computer Assisted Structure Evaluation) program, now elaborated to an enhanced form called MULTICASE [58], is an example of an algorithm which seeks biophores in the active and inactive compounds in a set. Input to the CASE program is by means of a line notation system, called KLN, which has similarities to the Wiswesser line notation system (see ref [8] for a discussion of line notation systems). The program generates a very large number of descriptors, as one report states 'easily ranging in the thousands for 50–100 compound databases', consisting of molecular fragments of various sizes; molecular connectivity indices and log P are included in the MULTICASE program. Statistical tests are used to assess the significance of the biophores although, with such a large number of descriptors, the danger of chance effects cannot easily be overlooked. The CASE program has been applied to a variety of sets of biologically active compounds (ref. [58] and references therein) and, under the name CASETOX, to toxicity databases (see Section 9.2.2).

9.5   GENETIC METHODS

We have already seen how genetic methods can be applied to the search for multiple linear regression models (Section 6.3.1) and it has been briefly mentioned that they can be used in both the selection of neural network architectures and in their training. In principle, many different aspects of data analysis can benefit from a genetic algorithm 'wrapper' but there are also many other scientific tasks, involving selections from a large number of alternatives, which can be treated in this way. Thus, this section is a more general description of the use of genetic methods.

Genetic methods is a general term which covers genetic algorithms, evolutionary strategies, genetic programming and so on. There are subtle differences between these methods; genetic algorithms, for example, employ mating of fit solutions and also mutation of individual genes whereas evolutionary strategies make use of mutation, but the following general principles apply to all of these approaches.

The first step in a genetic approach is the representation of a solution which requires:

- choice of a coding scheme;
- choice of the form of the genetic vector;
- choice of the genetic alphabet.

There are 4 (or more) coding schemes; *gene-based* coding where each position in the vector represents an independent variable and each variable can assume a value independent of the others; *node-based* coding where each position represents a path; *delta* coding which is similar to gene-based except that the values of a particular solution are added to a 'template' solution and *messy* coding which is based on a template structure but here the position and value of the variables is important (missing values are allowed).

The genetic vector comes in a *standard form*, which contains values of the variables, and an *extended form* which is the same as standard but with extra control information. The genetic alphabet can use *real number*, *integer* or standard *binary* coding or a scheme called *grey coding* which is a variant of binary coding. Since the function/problem contains multiple minima, the coding scheme affects the 'distance' between minima and hence how easy it is to find a solution. Real number coding allows solution to any degree of accuracy – but there are infinite solutions.

Having chosen how to represent a solution there are a number of steps involved in the genetic process:

1. Determine the size of the *population* (the number of solutions to store).

2. Create an *initial population* (there are a variety of ways to do this, e.g. random selection).
3. Choose how to select the *parents* (again, various ways, e.g. random selection, based on *fitness*, etc.).
4. Determine a *mating operator* (the standard is a 1 point crossover) and the *probability of mating*.
5. Choose a *mutation operator* and *mutation probability*.
6. Decide whether to use a *maturation operator*. This can improve the search results by optimization of a solution before checking its fitness.
7. Choose which *offspring* to keep (again there are various ways, e.g. based on fitness (an *elitist* strategy) or simplicity, or a combination of the two, etc.).
8. Choose what to do with the new offspring. They may be added to the population by replacement of the weakest (least fit) or by replacement of random members of the population, etc.

Inherent in this process is some means of judging the fitness of a solution. In other words, how well a model has been fitted, how well a neural network performs, how closely a structure fits to a template and so on. Choice of a fitness function is crucial since it is the optimization of this which is the driving force of the genetic method.

Genetic methods have been applied successfully to a very wide range of scientific problems but they also suffer from some disadvantages:

- The configuration of the G.A. (coding scheme, fitness function, etc.) is crucial for success.
  - There are few guidelines to help with configuration.
  - Most configurations are problem dependent.
- They may reach a local optimum too quickly.
- They may suffer from slow convergence.

9.6 CONSENSUS MODELS

We have already seen the use of consensus models in Section 9.3.3 where the 100 'best' artificial neural networks from a set of 1000 were chosen to form a committee to make predictions of activity against kinase targets. This, of course, makes sense as any individual trained network will have converged to some minimum error value and may not be an optimum solution. Use of a panel of trained networks like this is likely to give a more reliable prediction and has the added advantage of being able to assign a range of likely values thus giving some measure of reliability. Another example of a situation where consensus models could be used is in the case of multiple small (3 or 4 term) multiple linear regression models derived from a large pool of potential independent variables. Just such a situation was shown in Section 6.3.1.5 where the starting set consisted of 53 variables. The use of different genetic strategies gave rise to populations of linear regression models and these could be combined to give consensus predictions. Some papers have reported improvements in predictions using consensus models of MLR equations, and of other types of model, but it has also been shown that in some situations there is no improvement and that the benefits don't outweigh the complexity [59]. These findings may be problem specific, of course, but further work is needed to draw conclusions on the utility of consensus models of the same type.

Another use of consensus modelling, however, is to combine models of different types. It should be obvious from a number of the examples shown in this book that it is possible to apply several different data modelling techniques to the same set of data. These usually, but not always, give rise to similar sets of predictions and in some cases may work better for one part of the data set, say the inactives, than another. Consensus modelling in these circumstances simply consists of running each of the available models on new data points to be predicted and then forming a consensus forecast in some way. In the case of classified data this may be just a majority vote on the outcome, for continuous data it may involve a range of predicted values. An example of the classification of androgenicity (compounds which mimic androgenic hormones such as testosterone) for a diverse set of molecules using three different classification techniques showed that a consensus model performed better than any of the individual models [60]. Like the regression results this finding may be problem specific but it is tempting to believe that a consensus made up from different modelling approaches may well give superior predictions.

9.7 SUMMARY

Sections 9.2.1 to 9.2.3 described a number of important chemical expert systems which are regularly used in the research and development of pharmaceuticals and agrochemicals. Hopefully, readers with other research or commercial interests may see an application of these and similar systems to their own work, or may even be prompted to develop new systems tailored to their own needs. Section 9.3 discussed neural networks which are a fascinating new development in artificial intelligence techniques and which appear to offer a novel method for the display of multidimensional data. They have been extensively applied to the analysis of different types of data sets and techniques have been devised to unravel the complexities of the non-linear models that they fit to data. Rule induction and the pharmacophore/biophore recognition systems appear to show promise in data analysis and should be useful complements to the wide range of more or less well-understood methods available today. Finally, genetic algorithms offer an attractive method for the optimization of complex problems and consensus modelling may well be an important step forward in the production of robust and reliable predictive models.

In this chapter the following points were covered:

1. how expert systems work;
2. the development of expert systems for chemical property and toxicity prediction;
3. the use of expert systems for chemical reaction planning and the prediction of chemical structures;
4. how neural networks work and how they are used in data display and data analysis;
5. attempts to uncover the 'black box' of artificial neural network models;
6. what is meant by rule induction;
7. how genetic algorithms can be set up to solve complex problems;
8. the possibility that consensus models may work better than individual models.

REFERENCES

[1] Ayscough, P.B., Chinnick, S.J., Dybowski, R., and Edwards, P. (1987). *Chemistry and Industry*, **Aug.**, 515–20.
[2] Cartwright, H.M. (1993). *Applications of Artificial Intelligence in Chemistry*. Oxford University Press, Oxford, UK.
[3] Jakus, V. (1992). *Collection of Czechoslovak Chemical Communications*, **57**, 2413–51.
[4] Nys, G.C. and Rekker, R.F. (1973). *European Journal of Medicinal Chemistry*, **8**, 521–35.
[5] Rekker, R.F. and de Kort H.B. (1979). *European Journal of Medicinal Chemistry*, **14**, 479–88.

[6] Hansch, C. and Leo, A.J. (1979). *Substituent Constants for Correlation Analysis in Chemistry and Biology*, pp. 18–43. John Wiley & Sons, Inc., New York.

[7] Mayer, J.M., Van de Waterbeemd, H., and Testa, B. (1982). *European Journal of Medicinal Chemistry*, **17**, 17–25.

[8] Weininger, D. and Weininger, J.L. (1990). In *Quantitative Drug Design* (ed. C.A. Ramsden), Vol. 4 of *Comprehensive Medicinal Chemistry. The Rational Design, Mechanistic Study and Therapeutic Application of Chemical Compounds*, C. Hansch, P.G. Sammes, and J.B. Taylor (eds), pp. 59–82. Pergamon Press, Oxford.

[9] Ashby, J. (1985). *Environmental Mutagenesis*, **7**, 919–21.

[10] Tennant, R.W. and Ashby, J. (1991). *Mutation Research*, **257**, 209–27.

[11] Ashby, J. and Tennant, R.W. (1991). *Mutation Research*, **257**, 229–306.

[12] Sanderson, D.M. and Earnshaw, C.G. (1991). *Human and Experimental Toxicology*, **10**, 261–73.

[13] Judson, P.N. (1992). *Pesticide Science*, **36**, 155–60.

[14] Langowski, J. (1993). *Pharmaceutical Manufacturing International*, 77–80.

[15] Enslein, K., Blake, B.W., and Borgstedt, H.H. (1990). *Mutagenesis*, **5**, 305–6.

[16] Enslein, K., Lander, T.R., Tomb, M.E., and Craig, P.N. (1989). *Toxicology and Industrial Health*, **5**, 265–387.

[17] Klopman, G. (1985). *Environmental Health Perspectives*, **61**, 269–74.

[18] Hileman, B. (1993). *Chemical and Engineering News*, 21 June, 35–7.

[19] Benigni, R. (2004). Prediction of human health endpoints: mutagenicity and carcinogenicity. In *Predicting Chemical Toxicity and Fate*, M. Cronin and D.J. Livingstone (eds), pp 173–92. CRC Press, Boca Raton.

[20] Corey, E.J., Long, A.K., and Rubenstein, S.D. (1985). *Science*, **228**, 408–18.

[21] Corey, E.J. (1991). *Angewandte Chemie – International edition in English*, **30**, 455–65

[22] Metivier, P., Gushurst, A.J., and Jorgensen, W.L. (1987). *Journal of Organic Chemistry*, **52**, 3724–38.

[23] Manallack, D.T. and Livingstone, D.J. (1992). *Medicinal Chemistry Research*, **2**, 181–90.

[24] Salt, D.W., Yildiz, N., Livingstone, D.J., and Tinsley, C.J. (1992). *Pesticide Science*, **36**, 161–70.

[25] Manallack, D.T. and Livingstone, D.J. (1994). Neural Networks – A Tool for Drug Design. In *Advanced Computer-assisted Techniques in Drug Discovery*, H. Van de Waterbeemd (ed.), Vol 3 of *Methods and Principles in Medicinal Chemistry*, R. Mannhold, P. Krogsgaard-Larsen and H.Timmerman (eds), pp. 293–318. VCH, Weinheim.

[26] Zupan, J. and Gasteiger, J. (1991). *Analytica Chimica Acta*, **248**, 1–30.

[27] Burns, J.A. and Whitesides, G.M. (1993). *Chemical Reviews*, **93**, 2583–2601.

[28] Zupan, J. and Gasteiger, J. (1993). *Neural Networks for Chemists*. VCH, Cambridge.

[29] Livingstone, D.J. and Salt, D.W. (1995). Neural networks in the search for similarity and structure-activity. In *Molecular Similarity in Drug Design*, P. Dean (ed.), pp. 187–214, Blackie Academic and Professional, London, Glasgow.

[30] Livingstone, D.J. (ed.) (2008). *Artificial Neural Networks – Methods and Applications*, Vol. 458 of Methods in Molecular Biology, J.M. Walker, Series editor, Humana Press.

[31] Livingstone, D.J., Hesketh, G., and Clayworth, D. (1991). *Journal of Molecular Graphics*, **9**, 115–18.

[32] Selwood, D.L, Livingstone, D.J., Comley, J.C.W., *et al.* (1990). *Journal of Medicinal Chemistry*, **33**, 136–42.

[33] Hudson, B., Livingstone, D.J., and Rahr, E. (1989). *Journal of Computer-aided Molecular Design*, **3**, 55–65.

[34] Prakash, G. and Hodnett, E.M. (1978). *Journal of Medicinal Chemistry*, **21**, 369–73.

[35] Andrea, T.A. and Kalayeh, H. (1991). *Journal of Medicinal Chemistry*, **34**, 2824–36.

[36] Goodacre, R., Kell, D.B., and Bianchi, G. (1993). *Journal of the Science of Food and Agriculture*, **63**, 297–307.

[37] Livingstone, D.J. and Manallack, D.T. (1993). *Journal of Medicinal Chemistry*, **36**, 1295–7.

[38] Livingstone, D.J., Ford, M.G., Huuskonen, J.J. and Salt, D.W. (2001). *Journal of Computer-Aided Molecular Design*, **15**, 741–52.

[39] Tetko, I.V, Livingstone, D.J. and Luik, A.I. (1995). *Journal of Chemical Information and Computer Science*, **35**, 826–33.

[40] Manallack, D.T., Tehan, B.G., Gancia, E., *et al.* (2003). *Journal of Chemical Information and Computer Science*, **43**, 674–9.

[41] Helle, H.B. and Bhatt, A. (2002). *Petroleum Geoscience*, **8**, 109–18.

[42] Manallack, D.T., Pitt, W.R., Gancia, E., *et al.* (2002). *Journal of Chemical Information and Computer Science*, **42**, 1256–62.

[43] Livingstone, D.J., Evans, D.A., and Saunders, M.R. (1992). *Journal of the Chemical Society-Perkin Transactions II*, 1545–50.

[44] Tetko, I.V, Villa, A.E.P and Livingstone, D.J. (1996). *Journal of Chemical Information and Computer Science*, **36**, 794–803.

[45] Livingstone, D.J., Manallack, D.T. and Tetko, I.V. (1997). *Journal of Computer-aided Molecular Design*, **11**, 135–42.

[46] Setiono, R. and Liu, H. (1996). *IEEE Computer*, March 1996, 71–7.

[47] Taha, I. and Ghosh, J. (1996). *Intelligent Engineering Systems Through Artificial Neural Networks*, **6**, 23–8.

[48] Setiono, R. (2000). *IEEE Transactions of Neural Networks*, **11**, 512–19.

[49] Kamruzzaman, S.M. and Islam, Md. M. (2006). *International Journal of Information Technology*, **12**, 41–59.

[50] Livingstone, D.J., Browne, A., Crichton, R., Hudson, B.D., Whitley, D.C. and Ford, M.G. (2008). In *Artificial Neural Networks – Methods and Applications*, D.J. Livingstone (Ed.), Vol. 458 of Methods in Molecular Biology, J.M. Walker, Series editor, pp. 231–48, Humana Press.

[51] Towell, G. and Shavlik, J.W. (1993). *Machine Learning*, **31**, 71–101.

[52] Kohavi, R. and Quinlan, J. (2002). In *Handbook of Data Mining and Knowledge Discovery*, W. Klosgen and J.M. Zytkow (eds), pp. 267–76, Oxford University Press, New York.

[53] Hudson, B.D., Whitley, D.C., Browne, A. and Ford, M.G. (2005). *Croatia Chemica Acta*, **78**, 557–61.

[54] Lewis, D.F.V., Moereels, H., Lake, B.G., Ioannides, C., and Parke, D.V. (1994). *Drug Metabolism Reviews*, **26**, 261–85.

[55] Quinlan, J.R. (1986). *Machine Learning*, **1**, 81–106.

[56] A-Razzak, M. and Glen, R.C. (1992). *Journal of Computer-aided Molecular Design*, **6**, 349–83.

[57] Derde, M.-P., Buydens, L., Guns, C., Massart, D.L., and Hopke P.K. (1987). *Analytical Chemistry*, **59**, 1868–71.

[58] Klopman, G. (1992). *Quantitative Structure–Activity Relationships*, **11**, 176–84.

[59] Hewitt, M., Cronin, M.T.D., Madden, J.C., *et al.* (2007). *Journal of Chemical Information and Modeling*, **47**, 1460–8.

[60] Ji, L., Wang, X., Qin, L., Luo, S. and Wang, L. (2009). *QSAR & Combinatorial Science*, **28**, 542–550.

10

Molecular Design

Points covered in this chapter

- The need for molecular design
- Quantitative structure-activity and structure-property relationships
- Characterization of chemical structure by measured and calculated properties
- Application to mixtures

10.1 THE NEED FOR MOLECULAR DESIGN

Most, if not all, of what follows in this chapter can be applied to the design of any 'performance' chemical. What is meant here by a performance chemical is a molecule which exerts a specific effect or which has some particular property or set of properties which are essential for the product to function in the way that it is intended. In some applications the performance chemical may form the bulk of the product, in others it may only be a small percentage and in others it may be just one of several molecules which are all important for the eventual successful functioning of the product. This latter situation, of course, may be described as a mixture and some chemical products, especially those derived from natural materials, consist of mixtures of molecules whose exact composition may be unknown. Mixtures are such a special case that they need separate consideration as discussed in Section 10.5. There are many chemical products in everyday use which are perfectly satisfactory but where it may be desirable to change or modify the performance ingredient for a

A Practical Guide to Scientific Data Analysis   David Livingstone
© 2009 John Wiley & Sons, Ltd

variety of reasons. This may be economic, simply to reduce costs, say, or it may be necessary for legislative reasons, such as the introduction of the REACH legislation in the EU, or it may be commercial in order to succeed against a competitor's product, and so on. Whatever the reason, this situation calls for the application of molecular design.

The scientific approaches used in molecular design have been largely developed in the pharmaceutical and, perhaps to a lesser extent, agrochemical industries. This is mainly because these companies have a long history of very high expenditure on research and the enormous costs of the development of new drugs are widely known. Thus, the examples that follow are almost exclusively drawn from these fields but I hope that readers from other academic and industrial fields will see how these approaches may be applied to their own problems.

10.2   WHAT IS QSAR/QSPR?

The fact that different chemicals have different biological effects has been known for millennia; perhaps one of the earliest examples of a medicine was the use by the ancient Chinese of Ma Haung, which contains ephedrine, to treat asthma and hay fever. Table 10.1 lists some important biologically active materials derived from plants; no doubt most readers will be aware of other bioactive substances derived from plants.

Of course it was not until the science of chemistry had become sufficiently developed to assign structures to compounds that it became possible to begin to speculate on the cause of such biological properties. The ability to determine structure enabled early workers to establish structure–activity relationships (SAR), which are simply observations that a certain change in chemical structure has a certain effect on biological activity. As an example, molecules of the general formula shown in Figure 10.1 are active against the malaria parasite, *Plasmodium falciparum*. The effect of structural changes on the biological properties of derivatives of this compound are shown in Table 10.2, where the chemotherapeutic index is the ratio of maximum tolerated dose to minimum therapeutic dose.

Such relationships are empirical and are semi-quantitative in that the effect of changes in structure are represented as 'all or nothing' effects. In this example, replacement of oxygen by sulphur (compounds 8 and 3) results in a decrease in activity by a factor of 5, but that is all that can be said about that particular chemical change. In this case there

Table 10.1 Some examples of plant-derived compounds.

Artemisin	Antimalarial	Sweet wormwood (*Artemisia annua* L.)
Ascaridol	Anthelminthic	American Wormseed (*Chenopodium anthelminticum*)
Aspirin	Analgesic	Willow bark (*Salix* sp.)
Caffeine	Stimulant	Tea leaves and coffee beans
Digitalis	Antiarrythmic	Foxglove (*Digitalis purpurea*)
Ephedrine	Sympathomimetic	Ma Huang (*Ephedra sinica*)
Filicinic acid	Anthelminthic	Fern (*Aspidium filix-mas*)
Nicotine	Stimulant	Tobacco (*Nicotiana tabacum*)
Permethrin	Insecticide	Chrysanthemum
Quinine	Antimalarial	Cinchona bark (*Cinchona officinalis*)
Reserpine	Tranquilizer sedative	Fern (*Rauvolfia* spp.)
Strychnine	Central nervous system stimulant	Seeds (*Strychnos nux-vomica*)
Taxol	Antitumour	Pacific yew tree (*Taxus brevifolia*)
Vinblastin and Vincristine	Antitumour	Rosy periwinkle (*Catharanthus roseus*)

is only one example of that particular substitution and thus it is not possible to predict anything other than the fivefold change in activity. If the set of known examples contains a number of such changes then it would be possible to determine a mean effect for this substitution and also to assign a range of likely changes in activity for the purposes of prediction.

An SAR such as that shown here only applies to the set of compounds from which it is derived, the so-called 'training set' as discussed in Section 1.4 and Chapters 2 and 3. Although this might be seen as a disadvantage of structure-activity relationships, the same qualification also applies

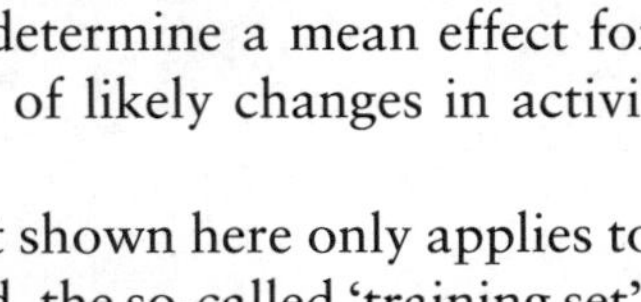

Figure 10.1 Parent structure of the antimalarial compounds in Table 10.2.

Table 10.2 Effect of structural variation on the antimalarial activity of derivatives of the parent compound shown in Figure 10.1.

	X	R_1	R_2	Chemotherapeutic index
1	$(CH_2)_2$	NO_2	OEt	0
2	$(CH_2)_2$	Cl	OMe	8
3	$(CH_2)_3$	Cl	OMe	15
4	$(CH_2)_3$	H	H	0
5	$(CH_2)_3$	Cl	OEt	7.5
6	$(CH_2)_4$	Cl	OEt	11.2
7	$(CH_2)_3$	CN	OMe	10
8	$(CH_2)_3$	Cl	SMe	2.8

to other quantitative models of the relationship between structure and activity. One of the powerful features of modelling is also one of its disadvantages, in that any model can only be as 'good' as the training set used to derive it. Making use of a number of more or less reasonable assumptions, the SAR approach has been used to derive more quantitative models of the relationship between structure and activity using a technique known as the Free and Wilson method which is described in Chapter 6.

Quantitative structure–activity relationships (QSAR) are a development of SAR. There is a similar term, QSPR, which is applied to relationships between a measured chemical property, e.g. solubility, boiling point, partition coefficient, etc., and chemical structure. So, what does this mean, what is QSAR/QSPR? The earliest expression of a quantitative relationship between activity and chemical structure was published by Crum Brown and Frazer in 1868 [1]:

$$\phi = f(C) \tag{10.1}$$

where ϕ is an expression of biological response and C is a measure of the 'constitution' of a compound. It was suggested that a chemical operation could be performed on a substance which would produce a known change in its constitution, ΔC. The effect of this change would be to produce a change in its physiological action, $\Delta\phi$. By application of this method to a sufficient number of substances it was hoped that it might be possible to determine what function ϕ is of C. It was recognized that the relationship might not be a strictly mathematical one because the terms ΔC, ϕ, and $\phi + \Delta\phi$ could not be expressed with 'sufficient definiteness to make them the subjects of calculation'. It was expected,

Table 10.3 Anaesthetic activity and hydrophobicity of a series of alcohols.

Alcohol	$\Sigma\pi$	Anaesthetic activity $\left(\log {}^1/_C\right)$
C_2H_5OH	1.0	0.481
$n-C_3H_7OH$	1.5	0.959
$n-C_4H_9OH$	2.0	1.523
$n-C_5H_{11}OH$	2.5	2.152
$n-C_7H_{15}OH$	3.5	3.420
$n-C_8H_{17}OH$	4.0	3.886
$n-C_9H_{19}OH$	4.5	4.602
$n-C_{10}H_{21}OH$	5.0	5.00
$n-C_{11}H_{23}OH$	5.5	5.301
$n-C_{12}H_{25}OH$	6.0	5.124

however, that it might be possible to obtain an approximate definition of f in Equation (10.1). The key to the difference between the philosophy of this approach and SAR lies in the use of the term quantitative. The Q in QSAR refers to the way in which chemical structures are described, using quantitative physicochemical descriptors. It does not refer to the use of quantitative measures of biological response, although this is a common misconception.

Perhaps the most famous examples of early QSAR are seen in the linear relationships between the narcotic action of organic compounds and their oil/water partition coefficients [2, 3]. Table 10.3 lists the anaesthetic activity of a series of alcohols along with a parameter, $\Sigma\pi$, which describes their partition properties (see Box 10.2 in this chapter for a description of π).

The relationship between this activity and the physicochemical descriptor can be expressed as a linear regression equation as shown below.

$$\log {}^1/_C = 1.039 \sum \pi - 0.442 \qquad (10.2)$$

Regression equations and the statistics which may be used to describe their 'goodness of fit', to a linear or other model, are explained in detail in Chapter 6. For the purposes of demonstrating this relationship it is sufficient to say that the values of the logarithm of a reciprocal concentration (log 1/C) in Equation (10.2) are obtained by multiplication of the $\Sigma\pi$ values by a coefficient (1.039) and the addition of a constant term (−0.442). The equation is shown in graphical form (Figure 10.2); the

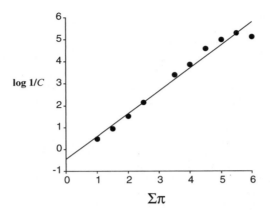

Figure 10.2   Plot of biological response (log 1/C) against $\Sigma\pi$ (from Table 10.3).

slope of the fitted line is equal to the regression coefficient (1.039) and the intercept of the line with the zero point of the x-axis is equal to the constant (−0.442).

The origins of modern QSAR may be traced to the work of Professor Corwin Hansch who in the early 1960s proposed that biological 'reactions' could be treated like chemical reactions by the techniques of physical organic chemistry [4]. Physical organic chemistry, pioneered by Hammett [5], had already made great progress in the quantitative description of substituent effects on organic reaction rates and equilibria. The best-studied and most well-characterized substituent property was the electronic effect, described by a substituent constant SIGMA (see Box 10.1).

Hansch, however, recognized the importance of partition effects in any attempt to describe the properties of compounds in a biological system. The reasoning behind this lay in the recognition that in order to exert an effect on a system, a compound first had to reach its site of action. Since biological systems are composed of a variety of more or less aqueous phases separated by membranes, measurement of partition coefficients in a suitable system of immiscible solvents might provide a simple chemical model of these partition steps in the biosystem.[1] Although the olive oil/water partition system had already been demonstrated to be of utility, Hansch chose octan-1-ol as the organic phase of his chemical model system of partition. Octan-1-ol was chosen for a variety of reasons: perhaps

---

[1] The organic phase of a partition coefficient system is intended to model the fatty, hydrophobic (water hating), membranes and the aqueous phase the hydrophilic parts of a biosystem.

Box 10.1   The electronic substituent constant, σ

Consider the ionization of benzoic acid as shown below where X is a substituent in the *meta* or *para* position to the carboxyl group.

The extent to which this equilibrium goes to the right, to produce the carboxylate anion and a proton, may be expressed by the value of the equilibrium constant, K_a^c, which is known as the concentration ionization constant

$$K_a^c = \frac{[A^-][H^+]}{[HA]}$$

where the terms in square brackets represent the molar concentrations of the ionized acid (A^-), protons (H^+), and the un-ionized acid (HA). This is a simplification of the treatment of ionization and equilibria but will serve for the purposes of this discussion. The 'strength' of an organic* acid, i.e. the extent to which it ionizes to produce protons, is given by the magnitude of K_a, most often expressed as the negative logarithm of K_a, pK_a. Since pK_a uses the negative log, a large value of K_a will lead to a small number and vice versa. Typical pK_a values of organic acids range from 0.5 (strong) for trifluoroacetic acid to 10 (very weak) for phenol. The strength of bases can also be expressed on the pK_a scale; here a large value of pK_a indicates a strong base. A very readable description of the definition and measurement of acid and base strengths, along with useful tabulations of data, is given in the monograph by Albert and Serjeant [6].

One of the features of an aromatic system, such as the benzene ring in benzoic acid, is its ability to delocalize electronic charge through the alternating single and double bonds. Once again, this is a simplification, since the bonds are all the same type; however, it will serve here. A substituent on the benzene ring is able to influence the ionization of the carboxyl group by donating or withdrawing electronic charge through the aromatic system. Since ionization produces the negatively charged

---

* Inorganic acids, such as HCl, H_2SO_4, and HNO_3, are effectively always completely dissociated in aqueous solution.

carboxylate anion, a substituent which is electron-donating will tend to disfavour this reaction and the equilibrium will be pushed to the left giving a weaker acid, compared with the unsubstituted, with a higher pK_a. An electron-withdrawing substituent, on the other hand, will tend to stabilize the anion since it will tend to 'spread' the negative charge and the equilibrium will be pushed to the right resulting in a stronger acid than the unsubstituted parent. Hammett [5] reasoned that the effect of a substituent on a reaction could be characterized by a substituent constant, for which he chose the symbol σ, and a reaction constant, ρ. Thus, for the ionization of benzoic acids the Hammett equation is written as

$$\rho\sigma_x = \log K_x - \log K_H$$

where the subscripts x and H refer to an x substituent and hydrogen (the parent) respectively. Measurement of the pK_a values of a series of substituted benzoic acids and comparison with the parent leads to a set of $\rho\sigma$ products. Choice of a value of ρ for a given reaction allows the extraction of σ values; Hammett chose the ionization of benzoic acids at 25 °C in aqueous solution as a standard since there was a large quantity of accurate data available. This reaction was given a ρ value of 1; the substituent σ values derived from these pK_a measurements have been successfully applied to the quantitative description of many other chemical equilibria and reactions.

the most important is that it consists of a long hydrocarbon chain with a relatively polar hydroxyl head group, and therefore mimics some of the lipid constituents of biological membranes. The octanol/water system has provided one of the most successful physicochemical descriptors used in QSAR, although arguments have been made in favour of other models and it has been proposed that three further chemical models of partition would be a useful addition to octanol [7]. It was suggested that these provide information that is complementary to that of the octanol/water system. When Hansch first published on the octanol/water system he defined [8] a substituent constant, π, in an analogous fashion to the Hammett σ constant (see Box 10.2).

The generalized form of what has now become known as the Hansch approach is shown in Equation (10.3).

$$\log {}^1\!/_C = a\pi + b\pi^2 + c\sigma + dE_s + \text{const.} \qquad (10.3)$$

where C is the dose required to produce a standard effect (see Section 10.3); π, σ, and E_s are hydrophobic, electronic, and steric parameters

Box 10.2 The hydrophobic substituent constant, π

The partition coefficient, P, is defined as the ratio of the concentrations of a compound in the two immiscible phases used in the partitioning system. The custom here is to take the concentration in the organic phase as the numerator; for most QSAR applications the organic phase is 1-octanol.

$$P = \frac{[\quad]_{OCT}}{[\quad]_{AQ}}$$

Here, the terms in the square brackets refer to the concentration of the same species in the two different phases.

Hansch chose logarithms of the partition coefficients of a series of substituted benzenes to define a substituent constant, π, thus

$$\pi_x = \log P_x - \log P_H$$

where x and H refer to an x-substituted benzene and the parent, benzene, respectively. The similarity with the Hammett equation may be seen but it should be noted that there is no reaction constant equivalent to the Hammett constant ρ. If a substituent has no effect on the partitioning properties of benzene, its π value will be zero. If it increases partition into the octanol phase, then P, and hence $\log P$, will be larger than for benzene and π will be positive. Such a substituent is said to be hydrophobic (water hating); a substituent which favours partition into the aqueous phase will have a negative π value and is said to be hydrophilic. Some representative π values are shown in the table.

Hydrophobic		Hydrophilic	
Substituent	π	Substituent	π
$-CH_3$	0.56	$-NO_2$	-0.28
$-C(CH_3)_3$	1.98	$-OH$	-0.67
$-C_6H_5$	1.96	$-CO_2H$	-0.32
$-C_6H_{11}$	2.51	$-NH_2$	-1.23
$-CF_3$	0.88	$-CHO$	-0.65

A couple of interesting facts emerged from early investigations of π values following the measurement of partition coefficients for several series of compounds. The substituent constant values were shown to be more or less constant and their effects to be, broadly speaking, additive. This is of particular importance if a quantitative relationship involving π is to be used predictively; in order to predict activity it is necessary to be able to predict values for the substituent parameters. Additivity breaks down when there are interactions between substituents, e.g. steric interactions, hydrogen bonds, electronic effects; 'constancy' breaks down when there is some interaction (usually electronic) between the substituent and the parent structure. One way to avoid any such problem is to use whole molecule log P values, or log P for a fragment of interest, but of course this raises the question of calculation for the purposes of prediction. Fortunately, log P values may also be calculated and there are a number of more or less empirical schemes available for this purpose [9, 10].

In the defining equation for the partition coefficient it was noted that the concentration terms referred to the concentration of the same species. This can have significance for the measurement of log P if some interaction (for example, dimerization) occurs predominantly in one phase, but is probably of most significance if the molecule contains an ionizable group. Since P refers to one species it is necessary to suppress ionization by the use of a suitable pH for the aqueous phase. An alternative is to measure a distribution coefficient, D, which involves the concentrations of both ionized and un-ionized species, and apply a correction factor based on the pK_a values of the group(s) involved. Yet another alternative is to use log D values themselves as a hydrophobic descriptor, although this may suffer from the disadvantage that it includes electronic information.

The measurement, calculation, and interpretation of hydrophobic parameters has been the subject of much debate. For further reading see references [10–12].

respectively (see Box 10.3); a, b, c, and d are coefficients fitted by regression; and const. is a constant. The squared term in π is included in an attempt to account for non-linear relationships in hydrophobicity. The form of an equation with a squared term is a parabola and it is true that a number of data sets *appear* to fit a parabolic relationship with the partition coefficient. However, a number of other non-linear relationships may also be fitted to such data sets and non-linear modelling in hydropobicity has received some attention, as described in Chapter 6.

What of QSPR? In principle, any property of a substance which is dependent on the chemical properties of one or more of its constituents could be modelled using the techniques of QSAR. Although most of the reported applications come from pharmaceutical and agrochemical research, publications from more diverse fields are increasingly beginning to appear. For example, Narvaez and co-workers [17] analysed the relationship between musk odourant properties and chemical structure for a

Box 10.3 The bulk substituent constant, MR

In Equation (10.3) three substituent parameters π, σ, and E_s, are used to describe the hydrophobic, electronic, and steric properties of substituents. The substituent constant E_s, due to Taft [13], is based on the measurement of rate constants for the acid hydrolysis of esters of the following type

$$X - CH_2COOR$$

where it is assumed that the size of the substituent X will affect the ease with which a transition state in the hydrolysis reaction is achieved.

A variety of successful correlations have been reported in which E_s has been involved but doubt has been expressed as to its suitability as a steric descriptor, mainly due to concern that electronic effects of substituents may predominantly control the rates of hydrolysis. Another problem with the E_s parameter is that many common substituents are unstable under the conditions of acid hydrolysis.

An alternative parameter for 'size', molar refractivity (MR), was suggested by Pauling and Pressman [14]. MR is described by the Lorentz–Lorenz equation

$$MR = \frac{n^2 - 1}{n^2 + 1} \cdot \frac{\text{mol.wt.}}{d}$$

where n is the refractive index, and d is the density of a compound, normally a liquid. MR is an additive-constitutive property and thus can be calculated by the addition of fragment values, from look-up tables, and nowadays by computer programs. This descriptor has been successfully employed in many QSAR reports although, as for E_s, debate continues as to precisely what chemical property it models. A variety of other parameters has been proposed for the description of steric/bulk effects [15] including various corrected atomic radii [16].

set of bicyclo- and tricyclo-benzenoids. A total of 47 chemical descriptors were generated (Table 10.4) for a training set of 148 compounds comprising 67 musks and 81 nonmusks. Using the final set of 14 parameters, a discriminant function (see Chapter 7) was generated which was able to classify correctly all of the training set compounds. A test set of 15 compounds, six musk and nine nonmusks, was used to check the predictive ability of the discriminant functions. This gave correct predictions for all of the musk compounds and eight of the nine nonmusks.

Table 10.4 Descriptors used in the analysis of musks (reproduced from ref. [17] with permission of Oxford University Press).

	Number of descriptors		
	Generated	Used[a]	Final[b]
Substructure	8	7	2
Substructure environment	6	6	2
Molecular connectivity	9	6	4
Geometric	15	8	4
Calculated log P	1	1	0
Molar refractivity	1	1	0
Electronic	7	6	2
Total	47	35	14

[a] Some descriptors were removed prior to the analysis due to correlations with other parameters or insufficient non-zero values.
[b] Number of parameters used in the final predictive equation.

Another example involves a quantitative description of the colour-fastness of azo dye analogues of the parent structure shown in Figure 10.3 [18]. Amongst other techniques, this study applied the method of Free and Wilson (see Chapter 6) to the prediction of colour-fastness. Briefly, the Free and Wilson method involves the calculation, using regression analysis, of the contribution that a substituent in a particular position makes to activity; here the activity is light-fastness of the dye. It is assumed that substituents make a constant contribution to the property of interest and that these contributions are additive. The analysis gave a regression equation which explained 92 % of the variance in the light-fastness data with a standard deviation of 0.49. An extract of some of the predictions made by the Free and Wilson analysis is shown in Table 10.5, which includes the best and worse predictions and also shows the range of the data. One advantage of this sort of treatment of the data is that it allows the identification of the most important positions of substitution (X_1 and X_5) and the most positively (CN and Cl) and negatively influential substituents (NO_2 and OCH_3).

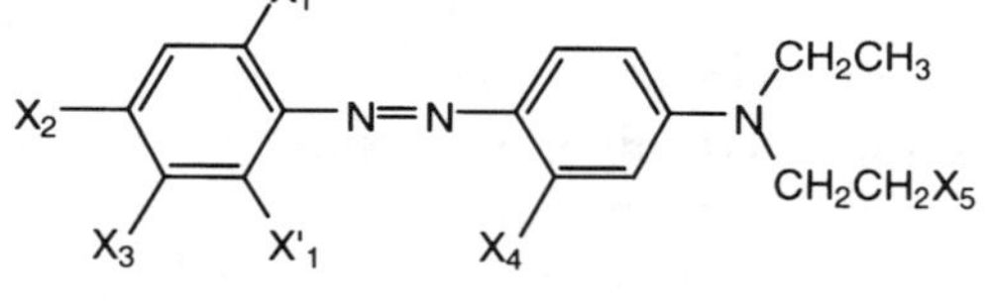

Figure 10.3 Parent structure of azo dye analogues (from ref. [18] copyright Society of Dyers and Colourists).

Table 10.5 Predicted light-fastness of azo dyes (from ref. [18] copyright Society of Dyers and Colourists).

Dye[a]	Calculated light-fastness	Residual[b]
1	4.03	0.47
4	5.50	0.00
8	5.35	−0.35
13	2.84	0.16
18	1.05	−0.05
19	6.69	0.31
21	4.25	0.75
28	2.15	−0.15
39	5.76	−0.76
44	5.36	−0.36

[a] Selected dyes from a larger set have been shown here.
[b] Difference between predicted and measured.

10.3 WHY LOOK FOR QUANTITATIVE RELATIONSHIPS?

The potential of organic chemistry for the production of new compounds is enormous, whether they be intended for pharmaceutical or agrochemical applications, fragrances, flavourings, or foods. In May 2009, *Chemical Abstracts* listed more than 46 million compounds, but this is only a tiny percentage of those that could be made. As an example, Hansch and Leo [19] chose a set of 166 substituents to group into various categories according to their properties (see Chapter 2). If we consider the possible substitution positions on the carbon atoms of a relatively simple compound such as quinoline (Figure 10.4), there are 10^{15} different analogues that can be made using these substituents. If the hunt for new products merely involved the synthesis and testing of new compounds without any other guidance, then it would clearly be a long and expensive task.

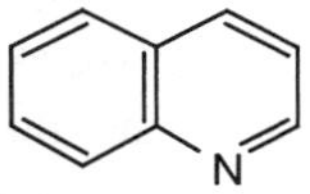

$$166^7 = 3.47 \times 10^{15}$$

Figure 10.4 Quinoline.

Of course, this is not the way that industry goes about the job. A large body of knowledge exists ranging from empirical structure–activity relationships to a detailed knowledge of mechanism, including metabolism and elimination in some cases. The purpose of quantitative structure–activity (or property) relationships is to provide a better description of chemical structure and perhaps some information concerning mechanism. The advantage of having a better description of structure is that it may be possible to transfer information from one series to another. In the example shown in Section 10.2, it was seen that substitution of a sulphur atom by oxygen resulted in an improvement in activity. This may be due to a change in lipophilicity, bulk, or electronic properties. If we know which parameters are important then we can, within the constraints of organic chemistry, design molecules which have the desired properties by making changes which are more significant than swapping oxygen for sulphur.

The work of Hansch et al. [20] provides an example of the use of QSAR to give information concerning mechanism. They demonstrated the following relationship for a set of esters binding to the enzyme papain.

$$\log {}^1\!/_{K_m} = 1.03\pi'_3 + 0.57\sigma + 0.61MR_4 + 3.8$$
$$n = 25 \qquad r = 0.907 \qquad s = 0.208 \tag{10.4}$$

Where K_m, the Michaelis-Menten constant, is the substrate concentration at which the velocity of the reaction is half maximal. The subscripts to the physicochemical parameters indicate substituent positions. The statistics quoted are the number of compounds in the data set (n), the correlation coefficient (r) which is a measure of goodness of fit, and the standard error of the fit (s); see Chapter 6 for an explanation of these statistics. It is possible to try to assign some chemical 'meaning' to the physicochemical parameters involved in Equation (10.4). The positive coefficient for σ implies that electron-withdrawing substituents favour formation of the enzyme–substrate complex. Since the mechanism of action of papain involves the electron-rich SH group of a cysteine residue, this appears to be consistent. The molar refractivity term (see Box 10.3) is also positive, implying that bulkier substituents in the 4 position favour binding. The two parameters π_4 and MR_4 are reasonably orthogonal for the set of 25 compounds used to generate Equation (10.4), and since the data does not correlate with π_4 it was concluded that a bulk effect rather than a hydrophobic effect was important at position 4. The prime sign associated with the π parameter for position 3 indicates that where there were two meta substituents the π value of the more hydrophobic substituent was used, the other π_3 value being ignored. The rationale for this procedure was that binding of one *meta* substituent to the enzyme placed the other *meta* substituent into an aqueous region outside the enzyme binding site. It was also necessary to make this assumption in order to generate a reasonable regression equation which described the data.

Following the QSAR analysis, Hansch and Blaney [21] constructed computer a model of the enzyme and demonstrated that the invariant hydrophobic portion of the molecules could bind to a large hydrophobic pocket. In this model, one of the two *meta* substituents also fell into a hydrophobic pocket forcing the other *meta* substituent out of the binding site. The substituent at the 4 position points towards an amide group on the enzyme which is consistent with the assignment of a bulk not hydrophobic component to enzyme binding at this position. The QSAR equation and molecular graphics study in this instance appear to tie together very nicely and it is tempting to expect (or hope!) that this will always be the case. A note of caution should be sounded here in that strictly speaking a correlation does not imply causality. However, there is no need to be unduly pessimistic; correlation can inspire imagination!

10.4   MODELLING CHEMISTRY

In the late 1970s there were two main approaches to molecular design: the techniques of QSAR as described in the previous section and elsewhere in this book and the use of molecular models, previously physical models but by this time computational. These were viewed as alternatives and each method had their champions and detractors. In truth, of course, they were actually complementary and gradually this realization dawned on everyone concerned.

A major problem with the QSAR approach was the description of molecular structure. The most information rich descriptors were the substituent constants as described in Boxes 10.1 to 10.3 but there were a number of drawbacks in their use:

- They could only be applied to congeneric series (that is, derivatives of a common parent).
- There were often missing values in the tabulations which could only be replaced by experimental measurements.
- For complex molecules it was sometimes difficult to decide what the common parent should be, and hence which series of substituent constants to use or which positional variants to use.

Box 10.4   Molecular connectivity indices

Molecular connectivity is a topological descriptor, that is to say it is calculated from a two-dimensional representation of chemical structure. All that is required in order to calculate molecular connectivity indices for a compound is knowledge of the nature of its constituent atoms (usually just the heavy atoms, not hydrogens) and the way that they are joined to one another.

Consider the hydrogen-suppressed graph of the alcohol shown below.

The numbers in brackets give the degree of connectivity, δ_i, for each atom; this is just the number of other atoms connected to an atom. For each bond in the structure, a bond connectivity, C_k, can be calculated by taking the reciprocal of the square root of the product of the connectivities of the atoms at either end of the bond. For example, the bond connectivity for the first carbon–carbon bond (from the left) in the structure is

$$C_1 = 1 \Big/ \sqrt{(1 \times 3)}$$

More generally the bond connectivity of the kth bond is given by

$$C_k = 1 \Big/ \sqrt{(\delta_i \delta_j)}$$

where the subscripts i and j refer to the atoms at either end of the bond. The molecular connectivity index, χ, for a molecule is found by summation of the bond connectivities over all of its N bonds.

$$\chi = \sum_{k=1}^{N} C_k$$

For the butanol shown above, the four bond connectivities are the reciprocal square roots of (1×3), (1×3), (2×3), and (2×1) which gives a molecular connectivity value of 2.269. This simple connectivity index is known as the first-order index because it considers only individual bonds, in other words paths of two atoms in the structure. Higher order indices may be generated by the consideration of longer paths in a molecule and other refinements have been considered, such as valence connectivity values, path, cluster, and chain connectivities [22].

Molecular connectivity indices have the advantage that they can be readily and rapidly calculated from a minimal description of chemical structure. As might be expected from their method of calculation they contain primarily steric information, although it is claimed that certain indices, particularly valence connectivities, also contain electronic information. Molecular connectivity has been shown to correlate with chemical properties such as water solubility, boiling point, partition coefficient, and Van der Waals' volume. Other topological descriptors have been used to describe a variety of biological properties including toxicity, and they have a number of environmental applications. There are many different types of topological descriptors which are well described in the book by Devillers and Balaban [23].

There was a class of descriptor available, however, which overcame all of these problems and these were the topological descriptors or molecular connectivity indices (see Box 10.4). There was some resistance to their use, mainly on the grounds of interpretation, but they were quite popular particularly in applications to environmental data which often involved diverse sets of compounds.

The major breakthrough in the description of molecules came about through the use of the computational molecular modelling packages. The first parameters calculated in this way were simple spatial descriptors based on the computational equivalent of physical models; quantities such as length, breadth, width, volume etc. Added to these were properties calculated from the results of semi-empirical, and sometimes *ab initio*, quantum mechanical calculations; quantities such as the energy of the highest occupied molecular orbital, the dipole moment and it's X, Y and Z components, atomic charges, superdelocalizability and so on. Research into the utility of different types of molecular descriptors led to an explosion in their numbers such that a handbook of molecular descriptors published in 2000 listed over 3000 different types [24]. Despite this, it seems that there is no generally accepted best set of parameters to use in quantitative molecular design [15].

10.5 MOLECULAR FIELD AND SURFACE DESCRIPTORS

All of the properties described in the previous section, with the exception of molecular connectivity indices, can be more or less equated to quantities which could be measured by some suitable experiment. It might

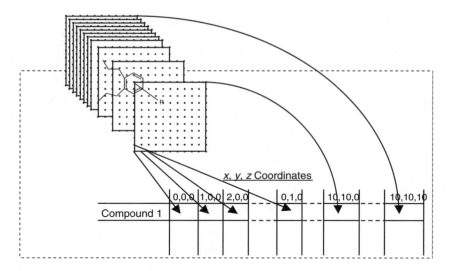

Figure 10.5 Illustration of the procedure for the production of a $10 \times 10 \times 10$ matrix of molecular field values (reproduced from ref. [15] copyright (2000) American Chemical Society).

be difficult or even impossible to devise experiments to measure some of the properties but in principle they are all descriptors which could be measured and which could be interpreted in physical or 'chemical' (e.g. in terms of reactivity, say) terms. Another class of property, known as molecular field descriptors, was devised in the late 1970s. This approach is based on the 3-D structure of a molecule and involves the superimposition of a grid of points in a box bounding the molecule. Probes are brought in to each of the grid points and an interaction energy calculated between the probe and the molecule at that point. Figure 10.5 illustrates this process.

The collection of grid point energies is known as a molecular field and different fields can be calculated using different probes and different interaction energy calculations, e.g. steric, electronic, hydrophobic and so on. The calculations are generally carried out using a molecular mechanics force field as described in Section 9.2.3 (Equation (9.2)). The whole process involves a number of steps:

- Obtain a suitable 3-D structure for each molecule in the training set.
- Derive partial atomic charges so that an electrostatic field can be generated.
- Align the molecules using some suitable alignment strategy (after conformational analysis if required).

- Create a cubic lattice of points around the molecules (usually larger than the largest member of the set).
- Compute interaction energies using a probe such as a pseudo methyl group with a unit positive charge. This generates a steric interaction energy based on a Lennard-Jones potential and an electrostatic interaction energy based on a coulombic potential.
- Fit a PLS model to the biological response and the interaction energies.
- Make predictions for a test set, visualize the results as contour plots on displays of the individual molecules in the set.

The advantages of this sort of description includes the fact that the 3-D structure of the molecules are involved and 3-D effects are known to be important in the interaction of drugs with biological systems. The first two systems to use this approach, CoMFA (Comparative Molecular Field Analysis) and Grid, have found many successful applications and a number of related techniques have subsequently been developed as described in reference [15].

A different sort of approach, but still based on 3-D structure, involves the calculation of molecular surfaces and then properties on those surfaces. These calculations involve quantum mechanics, not molecular mechanics, and are based on the surface of the molecules; not a field surrounding the structure. Early studies have shown promise for this technique [25].

10.6   MIXTURES

There are few reports on the application of quantitative design methods to mixtures. What has appeared has mostly been concerned with toxicity (e.g. [26], [27]) probably because of the importance of these effects and regulatory requirements. But mixtures are very important materials which we all use just about every day so why the paucity of effort in this area? There are, no doubt, a number of reasons but perhaps the most important lies in the difficulty of characterizing mixtures. Some approaches have used measured properties of the mixture and this can be useful in the development of empirical relationships with some other mixture property but it is not going to be predictive and is unlikely to be able to 'explain' the mixture property of interest.

So, how do we go about characterizing a mixture using properties which can be calculated and thus predicted for new components and/or mixtures? Consider the simplest mixture, a binary mixture of pure components in equal proportions (by mole fraction). Any property can be calculated for each of the components and the entire mixture may be characterized by a simple list of these properties thus:

$$P1_A, P2_A, P3_A, P4_A, \ldots\ldots, P1_B, P2_B, P3_B, \ldots\ldots$$

For mixtures of different mole fractions the properties can be weighted by the appropriate mole fraction in some way. The problem with this method, of course, is that it immediately doubles the number of descriptors that need to be considered and this can lead to problems of 'over-square' data matrices, that is to say data sets with more columns (descriptors) than rows (samples). An alternative is to only use parameters that are relevant to both components, i.e. whole molecule properties, and to combine these by taking the mole fraction weighted sum:

$$MD = R1 \times D1 + R2 \times D2$$

Where MD = Mixture descriptor, R1, R2 = mole fraction of first and second component in the mixture, D1, D2 = descriptor of first and second component. Application of this method to the density measurements of a very large set of binary mixtures led to some quite satisfactory models of deviation from ideal density as shown in Figure 10.6.

The results shown here are for consensus neural network models built using 15 calculated properties for a training set of nearly 3000 data points derived from 271 different binary mixtures. This technique could, of course, be extended to more complex mixtures.

A problem with this approach, though, is that it is difficult if not impossible to assign any mechanistic interpretation to the resulting models. The models can be used for prediction and this is fine if that is all that is required but the descriptors themselves relate to two or more molecules and the modeling process, using an ensemble of neural networks, is at best opaque. An alternative technique has been proposed in which the descriptors are based on mechanistic theories concerning the property to be modeled [29]. In this case the property concerned was infinite dilution activity coefficients which are the result of intermolecular interactions between two components in the mixture. Thus, mixture descriptors were formulated using different mixing rules based on thermodynamic principles (see reference for details). Attempts to build linear models for this data set using multiple linear regression and PLS failed so consensus neural network models were built using just 5 mixture descriptors. The

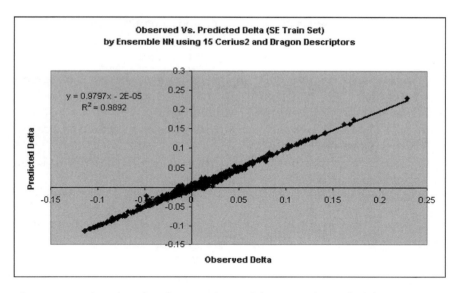

Figure 10.6 Plot of predicted versus observed deviations from ideal density (MED) using an ensemble neural network model (from ref. [28] copyright (2006) American Chemical Society).

importance of each of these parameters in the neural network models was judged by using a form of sensitivity analysis. This sensitivity analysis involved setting each descriptor one at a time to a constant value (its mean in the training set) and then calculation of the infinite dilution activity coefficients for the set using the neural network ensemble. The correlation coefficients for each of these models were compared with the correlation coefficient for the original model and the descriptors thus ranked in importance.

These are just two examples of how mixture properties may be modeled and no doubt, given the commercial importance of mixtures, other approaches will emerge in the future.

10.7 SUMMARY

The importance of molecular design has been described and some means for its implementation has been presented although the rest of this book contains many other examples. Approaches to the characterization of chemical structures have been briefly discussed, including some of their historical origins, and the difficulty of applying such methods to mixtures has been introduced. There is an enormous literature on this subject which the interested reader is encouraged to access.

In this chapter the following points were covered:

1. the reasons for molecular design and areas where it can be employed;
2. the meaning of the terms QSAR and QSPR;
3. how to characterize chemical structures using measured and calculated properties;
4. alternatives to the 'obvious' physical and chemical descriptors using fields and surfaces;
5. attempts to describe and/or explain the behaviour of mixtures.

REFERENCES

[1] Crum Brown, A., and Frazer, T. (1868–9). *Transactions of the Royal Society of Edinburgh*, **25**, 151–203.
[2] Meyer, H. (1899). *Archives of Experimental Pathology and Pharmakology*, **42**, 109–18.
[3] Overton, E. (1899). *Vierteljahrsschr. Naturforsch. Ges. Zurich*, **44**, 88–135.
[4] Hansch, C., Muir, R.M., Fujita, T., Maloney, P.P., Geiger, F., and Streich, M. (1963). *Journal of the American Chemical Society*, **85**, 2817–24.
[5] Hammett, L.P. (1937). *Journal of the American Chemical Society*, **59**, 96–103.
[6] Albert, A. and Serjeant, E.P. (1984). *The Determination of Ionization Constants: A Laboratory Manual* (3rd edn). Chapman & Hall, London.
[7] Leahy, D.E., Taylor, P.J., and Wait, A.R. (1989). *Quantitative Structure–Activity Relationships*, **8**, 17–31.
[8] Hansch, C., Maloney, P.P., Fujita, T., and Muir, R.M. (1962). *Nature*, **194**, 178–80.
[9] Livingstone, D.J. (1991). Quantitative structure–activity relationships. In *Similarity Models in Organic Chemistry, Biochemistry and Related Fields* (ed. R.I. Zalewski, T.M. Krygowski, and J. Shorter), pp. 557–627. Elsevier, Amsterdam.
[10] Livingstone, D.J. (2003). *Current Topics in Medicinal Chemistry*, **3**, 1171–92.
[11] Leo, A., Hansch, C., and Elkins, D. (1971). *Chemical Reviews*, **71**, 525–616.
[12] Dearden, J.C. and Bresnen, G.M. (1988). *Quantitative Structure–Activity Relationships*, **7**, 133–44.
[13] Taft, R.W. (1956). In *Steric Effects in Organic Chemistry* (ed. M.S. Newman), p. 556, Wiley, New York.
[14] Pauling, L. and Pressman, D. (1945). *Journal of the American Chemical Society*, **67**, 1003.
[15] Livingstone, D.J. (2000). *Journal of Chemical Information and Computer Science*, **40**, 195–209.
[16] Charton, M. (1991). The quantitative description of steric effects. In *Similarity models in organic chemistry, biochemistry and related fields*, (ed. R.I. Zalewski, T.M. Krygowski, and J. Shorter), pp. 629–87. Elsevier, Amsterdam.
[17] Narvaez, J.N., Lavine, B.K., and Jurs, P.C. (1986). *Chemical Senses*, **11**, 145–56.

[18] Carpignano, R., Savarino, P., Barni, E., Di Modica, G., and Papa, S.S. (1985). *Journal of the Society of Dyers & Colourists*, **101**, 270–6.

[19] Hansch, C. and Leo, A. (1979). *Substituent Constants for Correlation Analysis in Chemistry and Biology*. John Wiley & Sons, Inc., New York.

[20] Hansch, C., Smith, R.N., Rockoff, A., Calef, D.F., Jow, P.Y.C., and Fukunaga, J.Y. (1977). *Archives of Biochemistry and Biophysics*, **183**, 383–92.

[21] Hansch, C. and Blaney, J.M. (1984). In *Drug Design: Fact or Fantasy?* (ed. G. Jolles and K.R.H. Wooldridge) pp. 185–208. Academic Press, London.

[22] Kier, L.B. and Hall, L.H. (1986). *Molecular Connectivity in Structure–Activity Analysis*. John Wiley & Sons, Inc., New York.

[23] Devillers, J. and Balaban, A.T. (eds) (2000). *Topological Indices and Related Descriptors in QSAR and QSPR*, CRC, Boca Raton.

[24] Todeschini, R. and Consonni, V. (2000). *Handbook of Molecular Descriptors*, Wiley-VCH, Mannheim.

[25] Livingstone, D.J., Clark, T., Ford, M.G., Hudson, B.D. and Whitley, D.C. (2008). *SAR and QSAR in Environmental Research*, **19**, 285–302.

[26] Tichy, M., Cikrt, M., Roth, Z. and Rucki, M. (1998). *SAR and QSAR in Environmental Research*, **9**, 155–69.

[27] Zhang, L., Zhou, P.-J., Yang, F., and Wang, Z.-D. (2007). *Chemosphere*, **67**, 396–401.

[28] Ajmani, S.J., Rogers, S.C., Barley, M.H., and Livingstone, D.J. (2006). *J Chem Inf Model*, **46**, 2043–2055.

[29] Ajmani, S.J., Rogers, S.C., Barley, M.H., Burgess A.N., and Livingstone, D.J. (2008). *QSAR Comb. Sci.*, **27**, 1346–61.

Index

Note: Page numbers in *italics* refer to figures; those in **bold** to tables.